KV-370-665

Towards Gender Equity in Mathematics Education

An ICMI Study

Edited by

GILA HANNA
OISE,
Toronto, Ontario, Canada

KLUWER ACADEMIC PUBLISHERS
DORDRECHT / BOSTON / LONDON

A C.I.P. Catalogue record for this book is available from the Library of Congress.

ISBN 0-7923-3921-5 (Hb)
ISBN 0-7923-3922-3 (Pb)

Published by Kluwer Academic Publishers,
P.O. Box 17, 3300 AA Dordrecht, The Netherlands.

Kluwer Academic Publishers incorporates
the publishing programmes of
D. Reidel, Martinus Nijhoff, Dr W. Junk and MTP Press.

Sold and distributed in the U.S.A. and Canada
by Kluwer Academic Publishers,
101 Philip Drive, Norwell, MA 02061, U.S.A.

In all other countries, sold and distributed
by Kluwer Academic Publishers Group,
P.O. Box 322, 3300 AH Dordrecht, The Netherlands.

Printed on acid-free paper

TOWARD  ATICS EDUCATION

0144665

This book is due for return on or before the last date shown below.

Don Gresswell Ltd., London, N.21   Cat. No. 1208   DG 02242/71

079 233 9223

New ICMI Studies Series

VOLUME 3

Published under the auspices of The International Commission on Mathematical Instruction under the general editorship of

Miguel de Guzmán, President

Mogens Niss, Secretary

The titles published in this series are listed at the end of this volume.

CONTENTS

URSULA M. FRANKLIN CC, FRSC

PREFACE

THE REAL WORLD OF MATHEMATICS, SCIENCE, AND TECHNOLOGY EDUCATION

In this Preface, I would like to focus on what I mean by "education" and speak about the models and metaphors that are used when people talk, write, and act in the domain of education. We need to look at the assumptions and processes that the models and metaphors implicitly and explicitly contain. I feel we should explore whether there is a specific thrust to mathematics education in the here and now, and be very practical about it.

For me education is the enhancement of knowledge and understanding, and there is a strong and unbreakable link between the two. There seems little point in acquiring knowledge without understanding its meaning. Nor is it enough to gain a deep understanding of problems without gaining the appropriate knowledge to work for their solution. Thus knowledge and understanding are each necessary conditions for the process of education, but only when they are linked will the process bear fruit. Only in the balanced interplay of knowledge and understanding can we expect to achieve genuine education.

Fritz Schumacher, the author of *Small is Beautiful*, once spoke about a pun embedded in his name. In German, "schumacher" literally means the maker of shoes. But in order to do their work well — Schumacher reminded us — the makers of shoes needed to know about feet, about the activities and the lives of their customers, where the feet have to go, and what loads the people carry. The knowledge of how to make good shoes is truly useful only when linked to the lives of those who are to wear the shoes. This little meditation by Schumacher has always impressed me as a profound illustration of the link between knowledge and understanding — a link easily broken and difficult to restore.

Once we recognize the link between knowledge and understanding, we also recognize that knowledge is not neutral, objective, or value free. It is impossible to assume that science, technology, mathematics, or any other knowledge-seeking activity is neutral, because search, selection, and construction of new knowledge begins with questions — and questions arise in a given setting. Questions make sense only in a particular social and political context.

In her recent book *The Politics of Women's Biology*, Ruth Hubbard, a just-retired Harvard biologist, points out that scientists are the socially sanctioned fact-makers. However, scientists constitute a very small and homogeneous social group which in the past was almost entirely male, almost entirely white, and schooled in similar settings using similar or identical texts. Yet, as their insights and the results of their research become "facts," they shape the whole society. On the other hand, when those who work outside the in-group of scientists — say, women who nurse, cook, or garden — bring forward observations and insights, however well-tested and verified, these contributions rarely achieve the status of facts.

I raise these issues to point out that the knowledge we try to convey and the understanding we try to build is often very fragmented. The ways and means by which knowledge is accumulated and understanding is developed always structures the process of inquiry itself. Therefore it should not come as a surprise that dominant views on gender, race, and ideology have profoundly influenced scientific questions and scientific facts. In other words, the teaching of mathematics, science, and technology will be truncated and incomplete if it does not contain discussions about why certain problems are of interest and fundable at particular points in time while other questions don't seem to matter.

We also need to make it clear that experimental science in its reductionist and abstracting mode is but one source of understanding of the world around us. For instance, ecological problems show very clearly the limitation of traditional science as a basis for understanding and acting on some of the world's most urgent tasks.

Let me turn now to the process of education itself and to the metaphors we use to describe and discuss it. We often think about education as a natural process with underlying patterns. The oldest and most commonly recognized pattern is that of growth. Historically it was the natural cycle of growing, bearing fruit, maturity, decline, and decay that furnished metaphors for the interpretation of human and social activities.

Anyone who has planted a garden or brought up children has come to understand that growth can be nurtured and encouraged as well as hindered and stunted — but growth can not be commandeered. You cannot force carrots to grow — though you can plant them in the appropriate soil, water them, and care for them. Considerations of growth make it easier to accept the limitations of human intervention. However, I must make it quite clear at this point that bringing natural growth metaphors into our discussions of education does not mean accepting determinism. While carrots will be carrots, boys will not be boys in any but the most basic anatomical sense. What growth considerations do imply is recognition of and respect for the inner dynamics of development and maturation.

With the Industrial Revolution came a significantly different pattern of life — that of production. The new division of labour that took hold in 17th-century industrial Britain constituted a social invention of major consequence. In 1662, Sir William Petty, in his treatise on "The Wealth of England," pointed to the economic advantage of dividing the making of watches into a number of distinct steps, each the responsibility of a separate worker. One person would fashion the watch dials, one the handles, another the springs or the cases. Not only was the making of a watch less time consuming and costly, but workers in this new prescriptive production system needed to be familiar only with the sub-task assigned to them. They could be more easily trained and more readily replaced. Successful production depended not only on clear prescription and specification of the sub-tasks but also on effective management and planning. Much of the control of the work moved from workers to managers and planners. As the production mode of organizing work and people spread throughout industry as well as public and private administration, it furnished a powerful model for work and a new metaphor for social processes.

In today's society, our image of life and education is shaped far more strongly by a production model than by a growth model. Educators should be aware of the impact of the production model on their responsibilities. They should note that the demands for more and more frequent testing and evaluation, for detailed marking schemes and curriculum specifications are quality-control considerations, transferred from production experience to education. Such demands may set up serious conflicts with the insights flowing from a growth model of education. We cannot evade such conflicts but need to debate and mediate them.

Maybe we should now step back from models and metaphors and ask: "In terms of social purpose, what is the goal of education?" What would you answer? How does society benefit from having "educated" citizens? After all, in some discernible way, those who were "educated" by spending time in appropriate institutions ought to be different from those who did not.

Educators hope and often claim that educated persons will function more creatively, more productively, and more usefully within their society. Education is expected to show how acceptable contributions to the community at large can be made and what being a responsible member of the larger collectivity means. Assuming that such are the social goals of education, how does science and technology education fit into the picture? The core of our consideration of science and technology is not the question who is being educated, but why and how.

Though I am frequently very critical of applications of science, in particular technologies, it would not occur to me to say to anyone, "Forget about mathematics or electronics — this is stuff for manipulative uses." On the contrary, I advise especially those who are uneasy about a technology-dominated society to become competent enough to not only "read technology" but read between the lines, so that as citizens they can help to write a different technological text. But, as I see it, mathematics, science, and technology education has not yet found the vocabulary or the critical methodology that could be used for a textual analysis of science in a manner in which the use of language is analysed today. If technological literacy is one of our aims then such methods and discourses are urgently needed.

Another aspect of technological literacy requires attention. What textbooks are being used? What are our primers? I recall a conversation a number of years ago with a colleague, who had just negotiated with one of the big computer manufacturers a large gift of computers for his institution. Since we respected each other, I could say to him, "Please, just stop for a moment and take a deep breath and reflect. Think about the analogy of your computer gift and the gift of free Bibles." For the honourable purposes of literacy and education, many good people in the not-so-distant past donated Bibles to teach "heathens" to read. Of course, using the Bible as primer means more than just teaching how to read.

Thus the newly developed literacy brought with it acculturation that often resulted in the loss of indigenous culture. How different is the cultural programming of the computer from that of the cultural programming of the Bible? Should we not ask, How can we bring technological literacy to our students, using our text and primers without imposing undisclosed techno-values to the detriment or destruction of the students' own cultures and values? Rather than *not* teach for fear of indoctrination, we need to use many primers and explicitly transparent programming, and elucidate the purpose, strengths, and limitations of all devices or programs, and — most importantly — clarify the social assumptions that are embedded in every design.

It will take many pieces to complete the puzzle of how to educate in and for a world that is so decisively shaped by mathematics, science, and technology. We do have some of those pieces but many others still need to be invented or developed, which reminds me of the joke about the school principal who solemnly addressed an assembly by noting, "I see many who are missing." I too see much that is still missing and can only encourage all co-operative efforts to bring more pieces into the puzzle.

An analogy can help here. Imagine a society which is striving for universal literacy. Young children go to school to learn to read at an early age. Yet as they grow up, their teachers are not satisfied that they can read and comprehend a text or follow a set of written instructions. The responsible educator sees to it that the ability to read between the lines becomes part of the quest for literacy — because literacy encompasses not only the ability to read and understand the written text, but also to be mindful of what has not been expressed, though possibly implied.

Will those who advocate greater technological literacy today teach their students to understand technical instructions or will they teach them to read between the lines of what they experience of mathematics, science, and technology? Much benefit has come from the recent stress on "media literacy" — the ability to read and interpret mass-media offerings in terms of their institutional and political context. I think that we have not yet arrived at a similar level of "techno-literacy." There is not yet a public knowledge of mathematics, science, and technology that goes beyond the "gee-whiz" state. Yet such knowledge is needed to allow a critical understanding of some of the most powerful forces of our time.

There is then one more question for me to pose: "What do we do now —
how should we actually proceed in our own daily work in the field of mathematics, science, and technology education?" To me it seems important to
reaffirm, again and again, some very basic tenets of the work.

There is the need to reaffirm the primacy of people. Whatever we do, as
we teach children or adults, we must enrich the learners and must not cripple or incapacitate them. Thus, as technology becomes part of the education
system — be it as teaching instruments or as tools for skills transfer — the
impact on people has to be conscientiously assessed. Are such tools
enabling or disabling? Do they make individuals more resourceful, more
self-reliant, stronger, and more versatile? Or do they constrain human
potential? In particular, does the technological mediation of teaching and
learning constrain the imagination?

Are words not contained in the "spell check" unacceptable vocabulary?
Do they become unthinkable when a conventional expression is instantly at
hand? Sometimes I muse whether the Jabberwocky Bird could have been
created on a word processor and wonder whether "'Twas brillig ..." comes to
mind in front of a blank screen? If we rely heavily on tools that normalize
and standardize language then I feel we should find extra time to explore
non-standard language, to write poetry, rhyme, or compose nonsense verses.

Programming for learning must not become a cast that constrains imagination or creativity. Just as when a leg is put in a cast muscles can atrophy,
technological intervention in teaching and learning can atrophy mental
activities and hinder the leaps of the mind. Machines often free us from
routine and repetitive tasks — and we are ever grateful for not having to do
square roots or long division. Still, merely delegating such "simple" tasks
to devices can deprive the young learner of the joys of mastery and the
exploration of non-conventional solutions. While in traditional schooling
such learning occasions might occur naturally, technologically mediated
teaching will need to make special efforts to provide such opportunities.

This is particularly applicable to mathematics. Those of you who are
connoisseurs of mathematics will know how one rejoices in what can only
be called "an elegant solution." An elegant solution is not necessarily the
solution of an advanced problem but it is more than the correct solution of
any problem; it is a solution achieved with the minimum of fuss and the
maximum of ingenuity — elegant in its frugality, in its approach, in its
architecture. Even if we had the world's best programs to solve mathematical problems, I would contend that we must not, by their use in teaching,
deny people the joy of finding their own elegant solution through a combination of intellectual curiosity and serendipity.

Sometimes I tell a story about Gauss, the great German mathematician
(1977–1855). One day in the local one-room school the younger children
were given the task of adding the numbers from 1 to 100 on their slate,
while the teacher instructed the older students. The exercise was obviously
designed to keep the kids busy for some time. When Gauss came up with

the correct answer very quickly, he was accused of having learned the answer from his older brother who had done the same exercise the year before. Gauss, reportedly a somewhat withdrawn and sullen child, insisted that he had no help.The headmaster asked Gauss how he had found the solution so fast. The child pointed out that all he did was to add $100 + (99 + 1) + (98 + 2) +$ all the other blocks of 100 as well as the residual 50. Now there is an elegant solution that all could have found, though nobody else did.

Not all our students will turn out to be creative mathematicians, but the joy of the elegant solution (even to simple problems) ought not to be denied any learner young or mature. If the use of the devices eliminates those settings in which a Gauss-type intervention is possible, then equivalent settings should be in place *before* the devices become part of the curriculum.

To recap, then, I think that, as the education system continues to incorporate technology into the teaching environment, the primacy of people must be affirmed and re-affirmed. Technology, whether a tool of teaching or a subject of instruction, must enrich learning and human growth and not cripple or constrain either. This principle provides the measure for our ongoing critique.

Second, the joy of learning must not be stifled by the efficiency of teaching. No one should miss the excitement of the elegant solution, of seeing the hitherto unseen, of recognizing patterns in what appeared random — the sense of delight that I still have (after so many years of practice) when I look through a microscope.

Third, I think we must affirm the resourcefulness and resilience of people, and build teaching and learning on these attributes. Technology, not of necessity but often by design or implementation, is essentially anti-people. This was already evident at the beginning of the Industrial Revolution when factory owners would dream of machines to replace all workers. The worker-less factory totally under the owners' control became the desirata, perpetuating the notion that people were sources of problems, while machines and devices offered modes of solution.

As teachers and as citizens we must be careful to recognize, celebrate, and utilize the resourcefulness and resilience of people. Thus we must never lose the vision of education as a way for all teachers and learners alike: *to think and to be thoughtful*. In this light, education has always been a subversive activity. I urge you to keep it this way!

NOTE

This paper is based on an address given by Dr. Franklin at an invitational colloquium hosted by the MSTE Group, Faculty of Education, Queen's University, May 16 to 17, 1991. It has been reproduced with the permission of the author. (c) 1992 Ursula Franklin.

GILA HANNA

INTRODUCTION – TOWARDS GENDER EQUITY
IN MATHEMATICS EDUCATION

This volume of essays, *Towards Gender Equity in Mathematics Education*, has its origins in the October 1993 ICMI Conference, "Gender and Mathematics Education." About 100 mathematicians and mathematics educators from around the world gathered in Höör, Sweden, under the auspices of the International Commission on Mathematical Instruction, to discuss the well-documented world-wide gender imbalance in mathematics learning and mathematics-related careers. The Proceedings of this conference are being published under separate cover by the University of Lund Press. Both publications have been undertaken in the hope that they will encourage and inform the efforts being made in many countries to redress this imbalance.

The present volume, with one exception, consists of original articles solicited by the Editor to develop further the themes discussed at the ICMI conference. (Elizabeth Fennema's essay appears in the Conference Proceedings and is included here with the Editors' permission.) The premise of the ICMI Conference, that there is no physical or intellectual barrier to the participation of women in mathematics, science, and technology, provides the starting point for the analyses presented here. The authors explore the attitudinal and societal reasons for the gender imbalance in these fields and identify potential foci for change, among them curriculum and assessment practices, classroom and school cultures, and teacher education. A major portion of the book is devoted to detailed descriptions of educational systems around the world from the perspective of the role of gender in mathematics. These studies explore three interpretations of the concept of gender equity: (1) as equal opportunity to study mathematics; (2) as equality in mathematics experience in schools and classrooms; and (3) as equal educational outcomes.

Part One, General Issues, summarizes and assesses some of the research that has been conducted in the past two decades on gender equity in mathematics education. In "Mathematics, Gender, and Research," Elizabeth Fennema sets out the parameters for discussion. She reports: "in spite of some indications that achievement differences are becoming smaller — and they were never very large anyway — they still exist in those areas involving the most complex mathematical tasks, particularly as students progress to middle and secondary schools. There are also major differences in participation in mathematics-related careers. Many women, capable of learning the mathematics required, choose to limit their options by not studying mathematics.... Many

1

G. Hanna (ed.), Towards Gender Equity in Mathematics Education, 1–7.

of the differences that were reported in the 1970s, while smaller overall than they were then, still exist in 1994." Fennema speculates whether mathematics may be intrinsically biased, as part of the legacy of male-dominated society, and whether it thus makes sense to consider constructing a feminist mathematics.

In the next essay, "Gender and Mathematics: Mythology and Misogyny," Mary Gray suggests that for women the major barrier to achievement in mathematics and related fields is the mythology surrounding their aptitudes and abilities. Rejecting the argument made by some feminist mathematicians that mathematics must change if women are to do mathematics, she makes the charge that it is misogyny that keeps so many women away from the field and imposes a glass ceiling on those in it.

In "Gender Equity: A Reappraisal," Gilah Leder compares definitions of feminism commonly found in the research literature with the portrayal of gender issues in the popular print media. She then matches each of these viewpoints with models of mathematics programs familiar to educators, students, and parents. Leder chooses media reports as the "linking vehicle" in her reappraisal of gender equity, because of their considerable penetration into everyday culture and their resulting capacity to mould and reinforce popular opinion and expectations. By mapping feminist perspectives onto educational initiatives, she helps clarify their theoretical assumptions.

In her essay, "Symbolic Interactionism and Ethnomethodology as a Theoretical Framework for Research on Gender and Mathematics," Helga Jungwirth points out that the concepts developed in these two social science traditions have already proven useful in research on gender and mathematics. Their specific value for research on gender and mathematics is seen primarily in the emphasis they place on the active role of girls in mathematics classrooms and on the interactive creation of "facts." Jungwirth illustrates with classroom case studies how girls influence teacher-student interaction and how mathematical competence is itself interactively established.

The next essay, "Curriculum and Assessment: Hitting Girls Twice" by Sharleen Forbes, examines curriculum and assessment in New Zealand's secondary school system. The author criticizes the mathematics curriculum for reflecting male priorities (for using examples from the male world, among other things). Comparing "internal" assessment methods with traditional written examinations, Forbes also shows that females generally do better on the former and males on the latter. She maintains that females are being disadvantaged twice, by the curriculum and by assessment.

In the last essay in the section, "Mathematics and Gender: Some Cross-Cultural Observations," Ann Hibner Koblitz discusses some commonly held beliefs about the participation of women in mathematics and mathematics education, both historically and across cultures. She points out that their participation actually varies widely across both cultures and historical periods. Noting that women have been and are now actively involved in the world of mathematics in many countries of Asia, Africa, and Latin America as well as in Europe and North America, Koblitz examines essentialist notions of women's mathematical ability and interest, and offers a critique of certain feminist theories about the relationship between gender and science.

Part Two, Cross-Cultural Perspectives, provides an up-to-date appraisal of mathematics education and gender-equity initiatives in various countries. Barbro Grevholm opens this section by offering a detailed description of the Swedish system. Reviewing the representation of females in mathematics programs at various levels, including teacher education and university teaching, she points out that participation in mathematics programs seems to be a more serious problem in Sweden than in many other countries (though the performance of those who do participate is not a cause of concern). She notes that the seriousness of the situation is amplified by the lack of positive developments, and raises the question — given that the government supports change — of what the invisible barriers to women really are. Grevholm also outlines, based upon a recent report, some of the principal suggestions that have been put forward with a view to changing the status of women in mathematics education.

Next, Kari Hag provides us with a picture of mathematics education and gender in Norway. Noting that there has been very little focus on this subject in her country, she begins by drawing the gloomy picture first presented at a major conference in 1992. Hag provides an historical overview of the Norwegian school system from its cathedral school origins in the Middle Ages, then brings us up to date with a description of the current situation of females in education. She describes the initiatives of the Norwegian government to promote equity, reviews some studies on gender and mathematics, and then briefly outlines the new compulsory curriculum now being implemented as part of broader educational reforms.

In their review of the situation in Denmark, Bodil Branner, Lisbeth Fajstrup, and Hanne Kock point out that despite a formal commitment to equity, there remain large differences between men and women in education, career progress, salaries, and political involvement. The authors review the experiences of females as students and teachers throughout the educational system (including the popular universities, which are non-vocational institutions for adult education) with a particular focus on mathematics. The authors observe that gender polarization first becomes noticeable at the time of university entrance, when women seem to avoid the "hard" disciplines. They argue that there is a need, at all levels, to encourage women and to be more aware in particular of gender inequities among teachers. The authors believe that the situation will not be improved without special initiatives.

Lena Finne in her study of gender and mathematics education in Finland begins by acknowledging that there has been little research done on this topic in her country. She provides an extensive historical overview of Finland's public education system, dating back to its origins in 1809 as a merger of the Swedish and Russian systems. This is followed by specific data on the participation and achievements of girls and women in education generally and in mathematics programs in particular. Finne emphasizes that despite long-standing efforts in her country to create equitable opportunities for all, an achievement gap between boys and girls in mathematics persists and is particularly evident at the level of senior secondary school.

The next report, by Cornelia Niederdrenk-Felgner, examines the German situation. The author begins with an historical overview of education for boys and girls in Germany, with particular emphasis on mathematics education. She then presents data on the participation of women and girls which demonstrate that mathematics is still very much a male domain in Germany. She also reviews research on the role of gender in mathematics and on attitudes towards the topic, discussing in particular the teacher-student interaction as well as school textbooks and educational materials, reporting that discussions of this topic have become more intense in response to some recent initiatives and that mathematics educators now consider it an important topic to integrate into teacher and in-service training.

Josette Adda analyses the relevance of gender to mathematics education in France. She asserts that gender is not a relevant variable in the study of mathematics, but that it is relevant to career orientation and choice. The paper interprets and explains the imbalance between these two facts and concludes that it is a consequence of the conjunction of various prejudices of parents, teachers, employers, and students. This situation can only be changed, suggests the author, through information and knowledge. For girls, being successful in their mathematics studies is not enough; they also have to contend with many social constraints and prejudices.

Rosa Maria Spitaleri begins her review of gender and mathematics in Italy by pointing out that in her country more women than men attend university and that women are well represented in mathematics education and mathematics-related careers. In Italy teaching mathematics has traditionally been a woman's job, in fact. But teaching is generally undervalued in Italian culture, and so too are mathematics teachers. Statistics also show that at the highest levels of pure and applied mathematics and science women are under-represented. This leads the author to reflect on how the work of women, both formal and informal, is undervalued and how this undermines their self-esteem. Spitaleri is particularly concerned about the social and cultural barriers to professional advancement, and she urges women in mathematics-related fields to act collectively to promote the cause of gender equity.

Teresa Smart charts the progress made by girls in mathematics in England and Wales during the 1970s and how these gains were lost in the late 1980s and afterwards. She notes that the gender issue should not be seen in isolation, since the barriers which restrict black and working-class children often affect all girls. In her view, the earlier move towards equal opportunity was the result of an economic strategy on the part of the government that saw women primarily as skilled labour, particularly as scientists and engineers. Thus the recent economic recession and the reduced need to encourage women in mathematics and science allowed middle- and upper-class white men to reassert their dominance and take away the earlier gains of girls and women. Most striking, Smart observes, was the speed with which this was done and the inability of the educational establishment and of women to resist it.

Janice Gaffney and Judith Gill discuss mathematics education in Australian schools and the ways in which structures of tertiary education build on the gender divisions already established in schools. Aspects of gender and school mathematics dealt with include participation, achievement, and enrolment. The authors acknowledge that the situation has improved somewhat since the enactment of federal and state equity legislation over the past ten years, but point out that many projects, though well-intentioned, have not been particularly effective.

Megan Clark reviews schooling and educational opportunity in New Zealand. She outlines the goals, curriculum, and assessment in mathematics education and the participation and performance of boys and girls at the elementary and secondary school levels, where disturbing patterns of participation for young women appear by Year 12. She goes on to review the participation and outstanding achievement of women in some mathematical areas at the university and graduate levels, as well as the employment of women in academic roles in university mathematics and statistics departments. The author concludes that there are no particular concerns about women's mathematics performance in New Zealand, but that participation is an issue, especially in late secondary school, and that this period that should be targeted with a view to increasing the proportion of women in university mathematics studies. Clark also explores the situation of Maori women in senior secondary school in the past decade. She finds that their participation and performance in mathematics is an area of real concern, and that much needs to be done to retain them beyond the compulsory school years.

Rui Fen Tang, Qi Ming Zheng, and Shen Quan Wu provide us with "a snapshot of China" in terms of gender and mathematics education. During the thousands of years of feudal rule in China, there was no equality of the sexes, but there have been great changes since 1949, when a number of decrees were promulgated to improve the situation of women. The authors maintain, however, that traditional ideas about the place of women in education are still very much entrenched, citing current statistics which show that the proportion of female students and teachers drops off with each advance through the various levels of the education system. They also describe a secondary school experiment that attempted to improve the mathematics performance of girls through of a number of measures, among them the adoption by mathematics teachers of non-sexist attitudes, the cultivation of the girls' confidence in their ability to learn mathematics, the introduction of enhanced individual coaching for girls, and increased stress on the foundations of mathematics learning. Although the experiment involved one teacher at one school, the results were positive and encouraging. The authors propose a number of tasks which need to be undertaken to remove gender barriers to the achievement potential of girls in mathematics education in China.

The next essay takes us to Mexico, an underdeveloped country with a well-defined economic and cultural class system. Gender differences are strong, and the role of women is moulded by the value placed on the family and the expectation that the woman will serve as mother figure and source of family stability. Carlos Bosch and Maria Trigueros provide a brief review of the dramatic changes in education in Mexico over the past ten years, including the government support of mass education and the enormous growth of the university population. According to the authors, math-related disciplines have been slower than others to show broader participation. Although there is a lack of information about gender differences, it is widely accepted that it is predominantly men who pursue math-related careers. Some limited initiatives are under way to encourage girls and women in mathematics education and careers in face of the social pressure which still inhibits such a choice. The authors are optimistic about positive change as the education system continues to expand, expecting that future generations will be more receptive to gender equity.

Susan Chipman reviews female participation in the study of mathematics in the United States. She looks at mathematics at the university level, where, contrary to current belief, it is the least sex-typed of the major fields of study, and at the secondary school level, where the former large gender differences in enrolment have disappeared. Studies of US populations have shown little or no gender difference in overall mathematics performance prior to secondary school. The author reviews gender differences in mathematics enrolment and participation at the secondary school level and performance on standardized exams. At the graduate level, statistics indicate there may be a problem in female participation, the sources of which may warrant further investigation. In conclusion, Chipman observes that since the under-representation of US women in mathematics begins at the graduate level, it is incumbent upon mathematics departments to ask themselves whether their attitudes and advice to women are responsible.

It is apparent from these articles that the gender gap in mathematics study and achievement is both a complex reality for researchers and a difficult challenge for policy-makers and practitioners. We hope readers will find this volume an original useful contribution to this area of research and practice.

ACKNOWLEDGMENTS

I wish to thank the International Commission on Mathematical Instruction for initiating the Höör conference of October 1993 and for putting the role of gender in mathematics at the top of its agenda. Thanks are also due to the

International Organization of Women and Mathematics Education (IOWME) and to the Association for Women in Mathematics (AWM) for relentlessly pursing the goal of gender equity in mathematics education and for their very active participation at the conference. I also wish to thank Heather Berkeley, Donna Hutchins, and Doug Scott for their stylistic polishing of the papers and helpful advice in preparing this collection for publication. I also wish to thank Tobin MacIntosh for the book's design and typesetting.

ELIZABETH FENNEMA

MATHEMATICS, GENDER, AND RESEARCH

Research! What is it? Is it important? What knowledge about mathematics and gender does research contribute? And perhaps most importantly, can research provide new and different insights into the complex relationships between gender and mathematics? If we achieve this understanding, can we help females achieve equity in mathematics? Or are issues related to gender so value laden and embedded within our environment that an understanding of gender and mathematics can be attained only in relation to personal experience? Can change in mathematics education vis-à-vis gender only occur when society changes? The answers to these questions are complex, and probably they cannot be answered at all. However, I firmly believe that research can contribute to understanding gender and mathematics and that this understanding will help in achieving equity. The purpose of this paper is to report some of what we have learned from research and to speculate about what research can help us to learn in the future about the complexity of the interactions between gender and mathematics.

My own value positions influence strongly what I am going to say. This is nothing new. My entire professional career has been predicated on the belief that women deserve equity with men in all walks of life, and that belief has informed a significant part of my scholarly activities, particularly in the area of gender and mathematics. I have always believed that through research I can learn how to better facilitate the learning of mathematics by females.

I also believe that there are routes other than research for gaining knowledge about how to help females achieve equity. Many others have studied gender and mathematics and/or developed interventions designed to assist females in learning mathematics. Their work has greatly enriched my work, and I am indebted to them. However, this paper is focussed on what I know best — that portion of my professional work concerned with mathematics and gender.

RESEARCH, MATHEMATICS, AND GENDER

In 1974, my first article about gender was published in the *Journal of Research in Mathematics Education*. In this article, which was a review of extant work on sex differences in mathematics, I concluded that while many articles had been poorly analysed and/or included sexist interpreta-

G. Hanna (ed.), Towards Gender Equity in Mathematics Education, 9–26.

tions, there was evidence to support the idea that there were differences between girls' and boys' learning of mathematics, particularly in items that required complex reasoning; that the differences increased at about the onset of adolescence; and that these differences were recognized by many leading mathematics educators. As an aside, it was really the writing of that 1974 article that turned me into an active feminist, compelling me to recognize the bias that existed towards females, which was exemplified by the recognition and acceptance by the mathematics education community at large of gender differences in mathematics as legitimate.

The Fennema-Sherman studies (Fennema & Sherman, 1977, 1978; Sherman & Fennema, 1977), sponsored by the National Science Foundation and published in the mid 1970s, documented sex-related differences in achievement and participation in Grades 6 to 12. Although there were many subtle results, these findings basically agreed with those of my original review of gender differences in learning. In addition, Sherman and I found differences in the election of advanced level mathematics courses by males and females. When we coupled the achievement differences with the differential course enrolment, we hypothesized that if we could encourage females to participate in advanced mathematics classes at the same rate that males did, gender differences would disappear. Many things are learned as one does research, and from the stating of this hypothesis, I learned that what you write and say can stay with you a long time. This hypothesis, labelled as the *differential course-taking hypothesis*, became a point of attack by Julian Stanley and Camilla Benbow (Stanley & Benbow, 1980), who used their interpretations of some of their studies as a refutation of our hypothesis. They then used their work as evidence that gender differences in mathematics are genetic. Although widely attacked and disproved, the publication of their claims in the public media did have unfortunate repercussions (Jacobs & Eccles, 1985).

Affective or attitudinal variables were also examined in the Fennema-Sherman studies. Identified as critical were beliefs about usefulness of mathematics and confidence in learning mathematics, with males providing evidence that they were more confident about learning mathematics than were females, and males believing that mathematics was and would be more useful to them than did females. It also became clear that while young men did not strongly stereotype mathematics as a male domain, they did believe much more strongly than did young women that mathematics was more appropriate for males than for females. The importance of these variables, their long-term influence, and their differential impact on females and males was reconfirmed in many of our later studies, as well as by the work of many others (Leder, 1992).

One cognitive variable also studied in the Fennema-Sherman studies was spatial skills or spatial visualization, which I continued to investigate in a three-year longitudinal study in collaboration with Lindsay Tartre (Fennema & Tartre, 1985). Differences between females and males in spatial skills, particularly spatial visualization or the ability to visualize movements of geometric figures in one's mind, have long been reported. (Maccoby & Jacklin, 1974). Since items that measure spatial visualization are so logically related to mathematics, it has always appeared reasonable to believe that spatial skills contributed to gender differences in mathematics. We found that while spatial visualization is positively correlated with mathematics achievement (which does not indicate causation), not all girls are handicapped by inadequate spatial skills, but perhaps only those girls who score very low on spatial skills tests.

The Fennema-Sherman studies have had a major impact, although they were not particularly innovative, nor did they offer insights that others were not suggesting. However, they were published in highly accessible journals just when the concern with gender and mathematics was growing internationally. Partly because the studies were accessible and not generally controversial and because they employed fairly traditional methodology, their findings have been accepted by the community at large, and many have used them as guidelines for planning interventions and other research. The studies have been identified by two independent groups (Walberg & Haertel, 1992; Anonymous, in preparation) as among the most quoted social science and educational research studies during the last two decades. Each week, I still receive at least one request for information about the Fennema-Sherman Mathematics Attitude Scales, which were developed for those studies. The research reported in these studies, in conjunction with the research of others, has had a major impact. The problems of gender and mathematics were defined and documented in terms of the study of advanced mathematics courses, the learning of mathematics, and certain related variables that appeared relevant both to students' selection of courses and learning of mathematics.

After completing the Fennema-Sherman studies, with the indispensable aid of many others (Laurie Reyes Hart, Peter Kloosterman, Mary Koehler, Margaret Meyer, Penelope Peterson, and Lindsay Tartre), I broadened my area of investigation to include other educational variables, particularly teachers, classrooms, and classroom organizations. We studied teacher-student interactions, teacher and student behaviors, and characteristics of classrooms and teaching behaviors that had been believed to facilitate females' learning of mathematics.

The series of studies dealing with educational variables, reported and summarized in the book Gilah Leder and I edited (1990), suggested that it is relatively easy to identify differential teacher interactions with girls and boys: in particular, teachers interact more with boys than with girls, praise and scold boys more than girls, and call on boys more than girls. However, the impact of this differential treatment is unclear and difficult to ascertain. The data that resulted from the studies do not support the premise that differential teacher treatment of boys and girls causes gender differences in mathematics. This conclusion has also been reached by others (Eccles &

Blumenfeld, 1985; Koehler, 1990; Leder, 1982). As of 1994, there still is not sufficient evidence to allow us to conclude that interacting more or differently with girls than boys is a major contributor to the development of gender differences in mathematics.

Many intervention programs have been designed to help teachers recognize how they treat boys and girls differently. Unfortunately, such programs do not appear to have been successful in achieving the elimination of gender differences in mathematics. I believe that differential teacher treatment of boys and girls is merely a symptom of many other causes of gender differences in mathematics and that, as in medical practice, treating the symptom is not sufficient to change the underlying cause or condition.

Identifying behaviors in classrooms that influence gender differences in learning and patterns in how students elect to study mathematics has been difficult. Factors that many believed to be self-evident have not been shown to be particularly important. Consider sexist behaviors, such as those indicating that mathematics is more important for boys than for girls. No one would deny that such behaviors exist. However, Peterson and I (Fennema & Peterson, 1986; Peterson & Fennema, 1985) did not find major examples of overall sexist behaviors on the part of teachers, but rather small differences in teacher behavior, which, when combined with the organization of instruction, made up a pattern of classroom organization that appeared to favor males. We also found patterns of teacher behavior and classroom organization that influenced boys and girls differently. For example, competitive activities encouraged boys' learning and had a negative influence on girls' learning, while the opposite was true of co-operative learning. Since competitive activities were much more prevalent than co-operative activities, it appeared that classrooms we studied were more often favorable to boys' learning than to girls' learning.

In connection with this series of studies, Peterson and I proposed the Autonomous Learning Behaviors model, which suggested that because of societal influences (of which teachers and classrooms were main components) and personal belief systems (lowered confidence, attributional style, belief in usefulness), females do not participate in learning activities that enable them to become independent learners of mathematics (Fennema & Peterson, 1985). This model still appears valid, although my understanding of what independence is has grown, and I believe that independence in mathematical thinking may be learned through working in co-operation to solve mathematical problems.

INTERVENTION STUDIES

By about 1980, there were some rather consistent findings from research on gender and mathematics. We knew that in the United States, when studying mathematics became optional in the secondary schools, fewer females than males were electing to study mathematics; young women did not believe that mathematics was particularly useful and tended to have less confidence in themselves as learners of mathematics. We had strong evidence that boys stereotyped mathematics as a male domain and we had identified many societal influences that suggested that mathematics learning was not particularly appropriate for girls. Based on these research findings, with the help of three others (Joan Daniels Pedro, Patricia Wolleat, and Ann Becker DeVaney), I developed an intervention program called Multiplying Options and Subtracting Bias (Fennema, Wolleat, Becker, & Pedro, 1980). This approximately one-hour, school-based program, composed of videotapes and a workshop guide, was extensively evaluated, particularly with regard to its effectiveness in increasing girls' participation in advanced secondary school mathematics classes and its impact on confidence and perceived usefulness. We reported that this short intervention, which helped girls and boys recognize the importance of mathematics and the stereotyping of mathematics that was prevalent, resulted in more girls, and more boys, electing to take mathematics courses. A more negative finding, however, was that pointing out the sexism that exists in classrooms and their environment increased the girls' anxiety about mathematics.

Another educational finding, which has not been systematically studied by me, or anyone else that I know of, is the variation in gender differences in mathematics across schools and across teachers. Casserly (1980) reported that some schools were much more successful in attracting females to the most advanced mathematics classes than were other schools. The Fennema-Sherman studies reported variations in the size of the differences between schools. And one large urban school system in the US reported substantial discrepancies in gender difference scores among high schools on the SAT, a college entrance examination that is recognized as being a good test of mathematical reasoning (Personal Communication, 1985). Confirming this variation between schools and teachers, recent unpublished work of mine has identified certain teachers whose classes consistently produce scores with greater gender differences in favor of males than do the classes of other teachers.

My next set of studies was conducted with Janet Hyde. For these studies, we did a series of meta-analyses of extant work on gender differences reported in the US, Australia, and Canada (Hyde, Fennema, Lamon, 1990; Hyde, Fennema, Ryan, & Frost, 1990). The results indicated that while gender differences in mathematics achievement might be decreasing, they still existed in tasks that required functioning at high cognitive levels. It also seemed that the more nearly tests measured problem solving of the most complex cognitive level, the more tendency there was to find gender differences in mathematics in favor of males. The international assessment reported by Gila Hanna (Hanna, 1989) showed results that basically confirmed this.

My work has not been the sole chain of inquiry that has occurred during the last two decades. The work of Jacquelynne Eccles, Gilah Leder, and others (Leder, 1992) has been conducted independently and closely parallels the major themes I have addressed. An overly simplistic and not inclusive summary suggests that during this time, scholars documented that differential achievement and participation of females and males existed; next, some related educational and psychological variables were identified; explanatory models were then proposed; and finally (or concurrently in some cases) interventions, based on the identified variables, were designed to alleviate the documented differences.

One line of inquiry that I have not pursued, but which has added a significant dimension and more complexity to the study of gender and mathematics, is the work that has divided the universe of females into smaller groups. In particular, the work of the High School and Beyond Project (a large multi-year project that documented gender differences in mathematics as well as many other areas) as interpreted by Secada (1992) and the work of Reyes (Hart) and Stanic (1988) have investigated how socio-economic status and ethnicity interact with gender to influence mathematics learning. The US, as many other countries, is a highly heterogeneous society, made up of many layers, divisions, and cultures. The pattern of female differences in mathematics varies across these layers and must be considered.

For a number of years, because I was asked so often to speak about gender and mathematics, I compiled a list of what I had concluded that research, my own and others, had shown. Following is a portion of the list I made in 1990.

GENDER DIFFERENCES IN MATHEMATICS: 1990

1. Gender differences in mathematics may be decreasing.
2. Gender differences in mathematics still exist in:
    learning of complex mathematics
    personal beliefs in mathematics
    career choice that involves mathematics
3. Gender differences in mathematics vary:
    by socio-economic status and ethnicity
    by school
    by teacher
4. Teachers tend to structure their classrooms to favor male learning.
5. Interventions can achieve equity in mathematics.

On the basis of an examination of the five items on this list, which still reflect my thinking, it is clear that in the two decades following my original review in 1974, my understanding of gender and mathematics has grown as far as related variables are concerned, but the same gender differences, albeit perhaps smaller, still exist. I now can describe the problem more precisely. I know that large variations between groups of females exist; I know that there are differences among schools and teachers with respect to gender and mathematics issues; I know that females and males differ with respect to personal beliefs about mathematics; and I know that interventions can make a difference. I understand that the issue of gender and mathematics is extremely complex. And I accept without question the basic premise of the International Commission for Mathematics Instruction Study Conference on Gender and Mathematics (1992) — namely, that "there is no physical or intellectual barrier to the participation of women in mathematics." But in spite of all the work done by many dedicated educators, mathematicians, and others, the "problem" still exists in much the same form that it did in 1974.

Now before I sound too pessimistic, it should be noted that there are many females who are achieving in mathematics and are pursuing mathematics-related careers. However, let me reiterate that in spite of some indications that achievement differences are becoming smaller — and they were never very large anyway — they still exist in those areas involving the most complex mathematical tasks, particularly as students progress to middle and secondary schools. There are also major differences in participation in mathematics-related careers. Many women, capable of learning the mathematics required, choose to limit their options by not studying mathematics. And while I have no direct data, I strongly suspect that the learning and participation of many women, who might be in the lower two-thirds of the achievement distribution, have not progressed at all. I must conclude that many of the differences that were reported in the 1970s, while smaller overall than they were then, still exist in 1994.

NEW SCHOLARSHIP ON GENDER AND MATHEMATICS

Much of what we know about gender and mathematics has been derived from scholarship that has been conducted using a traditional social science research perspective — that is, a positivist approach that has basically looked at overt behaviors such as answers on a mathematics test, the amount one agrees with an item that is part of a confidence scale, interactions between a teacher and a student, or the career decisions that students make. Usually, in positivist educational research, studies are done by a researcher deciding which overt behaviors are important to study, figuring out a way to count or measure the behaviors, and then studying the counts in some way. Often groups are studied — for example, all females, all black females, or all females who study physics at university. Measures of central tendencies (averages) and variability between groups are compared, or relationships between the variables are examined. The purpose is usually to describe the groups with respect to a behavior, examine relationships between behaviors, or to use the results that are found to predict the behavior of others. For example, when the Fennema-Sherman studies reported differential achievement, the finding was based on comparing the mean responses and standard deviations on a standardized achievement test of a well-defined group of females with the mean responses and standard deviations of a group of similar males. Certain findings of differences were identified as statistically significant, which can be interpreted as meaning that the findings did not occur by chance but probably existed in the universe of similar males and females in the same proportion as in the population studied.

Studies conducted from a positivist perspective have provided powerful and rich information about gender and mathematics. Positivist scholarship should continue, particularly in order to continue the documentation of gender differences in participation and achievement in mathematics. However, an understanding of gender and mathematics based on studies done from this perspective is limited. Perhaps it is evidence of narrow vision, but I do not believe that we shall understand gender and mathematics until scholarly efforts conducted in a positivist framework are complemented with scholarly efforts that utilize other perspectives. Many educational researchers are using new perspectives, and their work is beginning to provide important insights into teaching, learning, and schooling. I believe that research conducted within these new perspectives would also provide important insights into the issues involving gender and mathematics.

Although there are many directions that scholarship on gender and mathematics could take, I would like to discuss and provide examples of research from two perspectives: cognitive science perspectives, which emphasize the irrelevance of female-male differences; and feminist perspectives, which emphasize that female-male differences are critical to the learning of mathematics.

COGNITIVE SCIENCE PERSPECTIVE

Brown and Borko (1992) define cognitive psychology as "the scientific study of mental events, primarily concerned with the contents of the human mind (knowledge, beliefs) and the mental processes in which people engage." Central to this perspective is the idea that much of behavior is guided by mental activity or cognitions. Since it is difficult to get at mental processes, data collection techniques and interpretation are very different from those used in positivist research. Usually the sample size is small, and individual interviews provide much of the data. In most such research related to schooling, researchers request that the subject report about his or her mental processes, either asking subjects to think aloud during problem solving or stimulating recall where subjects might view a videotape of their actions as they are asked to report what they were thinking. Researchers examine the personal reports and look for universals that apply to all people. Much of current research in the mainstream of mathematics education is being conducted utilizing a cognitive science perspective. Many studies of teachers' knowledge and beliefs, as well as most of the work on learners' thinking within specific mathematical domains, are examples of this approach.

Consider one of the more robust lines of inquiry in mathematics education, the work on addition and subtraction done with young children (Carpenter, Moser, & Romberg, 1982). Addition and subtraction were defined precisely as semantically different word problems that can be solved with addition/subtraction; children were asked in individual interviews to solve the problems, and the interviewer probed, either during or following problem solution, until he or she understood how the problems were solved. The researchers looked for and identified patterns of problem-solving behavior that seemed to reflect the mental activities of the children.

Out of this work, universals were identified about how young children, both females and males, come to understand basic arithmetic ideas. Counting and modelling-solution strategies, which are developed intuitively in order to make sense of one's environment, were among those universals. These universals are found in many cultures — both in those cultures where schooling is at a minimum and in highly schooled societies (Adetula, 1989; Olivier, Murray, & Human, 1990; Secada, 1991). These modelling and counting strategies provide the foundation for the child's development of more sophisticated understanding.

While most researchers working within this paradigm have not specifically investigated gender differences, for the last ten years Tom Carpenter and I have examined gender differences as we investigated the application to instruction of the universals identified by the work in addition and subtraction (Carpenter & Fennema, 1992). We have noted some differences in maturity between young girls and boys with respect to problem solving, but no differences in how boys and girls solve arithmetic problems. Thus, at our present level of knowledge, we believe that the processes used by females and males to make sense of arithmetic are essentially the same. And, when such findings are used by expert teachers, gender differences in mathematics appear to be irrelevant. Because such teachers can understand each individual's mental processing, instructional decisions do not appear to be influenced by teachers' beliefs related to gender. They teach to the cognitive level of their students and are not affected by superfluous issues.

Cognitive science research also has provided insights into teachers' behaviors, knowledge, and beliefs, although little has been done related to teachers' cognitions about gender. Such studies may lead to deeper understanding of gender differences in mathematics as understanding is gained about the mental life of students, teachers, and others, and how it influences daily decisions about learning mathematics. Unfortunately, there are not many studies related to gender that have been done using this perspective. Once again, I turn to my own and my colleagues' work. Our last study on gender and mathematics concerned teachers' knowledge of and beliefs about boys' and girls' successes in mathematics (Fennema, Peterson, Carpenter, & Lubinski, 1990). Although teachers thought the attributes of girls and boys who succeeded in mathematics were basically similar, teachers' knowledge about which boys were successful was more accurate than teachers' knowledge about which girls were successful, and teachers attributed the boys' successes more to ability and girls' successes more to effort. Linda Weisbeck's (1992) results add some interesting dimensions to our knowledge of teachers' cognitions. During stimulated-recall interviews, teachers reported that they thought more about boys than about girls during instruction. However, the characteristics they used to describe girls and boys were very similar.

It appears that teachers are very aware of whether the child they are interacting with is a boy or a girl. However, they don't think that there are important differences between girls and boys that should be attended to as they make instructional decisions. Boys just appear to be more salient in the teachers' minds. Teachers appear to react to pressure from students, and they get more pressure from boys. Interventions designed with this finding in mind would be very different from interventions that assume that teachers are sexist.

Another student, Carolyn Hopp (1994), who is currently finishing her dissertation, is concerned with what happens in co-operative small groups that influences the learning of mathematics, particularly the learning of complex mathematics like problem solving. While this work is still preliminary, it appears that boys and girls engage in different mental activities during co-operative problem-solving, and the impact on their learning of working in co-operative groups may be quite different depending on what mental activity is engaged in during the co-operative activity. Just working in small groups does not ensure that girls will learn mathematics. It depends on what goes on as the groups engage in co-operative activity.

Thus, while research conducted from a cognitive science perspective is still in its infancy as far as gender and mathematics are concerned, such studies can provide knowledge that will help us understand the underlying mechanisms that have resulted in gender differences in mathematics. Consider the case of the relationship between confidence in learning mathematics and the actual learning of mathematics. It has been assumed for at least two decades that lower confidence contributes to gender differences in mathematics. (In self-defense, if you read my writing carefully or listen to what I have said, you'll notice that I have never said that. In fact, I have often said that we do not know how confidence influences learning.) Perhaps a careful study of males' and females' perception of what has influenced their development of confidence in doing mathematics, and how their confidence has impacted on their study and learning of mathematics, might give us better insight into the relationship of the two. Or studying the impact that teachers' perception of the confidence of their students has on decisions that teachers make during mathematics instruction might provide deeper insight into teacher-student interactions.

Knowledge derived from a cognitive science perspective has enabled some teachers to eliminate gender differences in their mathematics classrooms. Carpenter and I have been investigating how knowledge of the universals of children's thinking about whole-number arithmetic could be used in classrooms and whether this knowledge would make a difference in what teachers did and how children learned (Carpenter & Fennema, 1992). We called the project Cognitively Guided Instruction (CGI) and we continue to investigate this today. Basically, we shared with teachers what we knew about the universals of children's learning, enabled them to become secure in that knowledge, and supported them as they applied the knowledge in their primary classrooms. Briefly, we found that teachers could acquire this knowledge of universals and use it in classrooms to make instructional decisions about individual children. CGI teachers' beliefs about children changed, and children in CGI classrooms have learned mathematics in excess of anything we expected.

At the beginning of our first study, before teachers had learned about children's thinking, we found that first-grade boys were better problem solvers than first-grade girls. In succeeding studies of children in Grades 1 to 3, who spent a year with teachers who know and understand children's thinking, we have found variable gender differences. Often, no differences exist; sometimes they are in boys' favor, and at other times they are in girls' favor. It appears that when teachers make instructional decisions based on their knowledge of individual children, overall gender differences are not found. It also appears that among certain teachers, although they are few in number, gender differences in favor of boys usually exist across classrooms and years; among even fewer teachers, differences in favor of girls are found across years. We are just beginning to ascertain whether we can identify components of their classrooms or cognitions that encourage the development of these gender differences in certain teachers' classrooms.

Thus, it appears that research utilizing a cognitive science perspective can be helpful in gaining an understanding of the relationship of gender to mathematics. It enables us to go beyond surface knowledge and overt behavior to develop an understanding of underlying mechanisms. With this understanding, future educational directions can be identified.

FEMINIST PERSPECTIVES

Included in the broad group of approaches to research that I am calling feminist are perspectives that have been defined as feminist methodologies, feminist science, feminist epistemologies, and feminist empiricism. I am not an expert on these and thus will not try to provide a thorough discussion. (For that, I refer you to Bleier, 1984; Campbell & Greenberg, 1993; Harding, 1987; and Shakeshaft, 1987). While within the work of scholars included in this tradition there are marked differences that go beyond the purview of this paper, these scholars do share a commonality. Without exception, they focus on interpreting the world and its components from a feminine point of view, and the resulting interpretations are dramatically different from what persists today.

Feminist scholars argue very convincingly that most of our beliefs, perceptions, and scholarship, including most of our scientific methodologies and findings, are dominated by male perspectives or interpreted through masculine eyes. According to feminist scholars, the view of the world as interpreted solely through masculine perspectives is incomplete at best and often wrong. If women's actions and points of view had been considered over the last few centuries, according to many of these feminist scholars, our perceptions of life would be much different today.

A basic assumption of feminist work is that there are basic differences between females and males that are more prevalent than the obvious biological ones and that result in males and females interpreting the world differently. Although many of these scholars present convincing arguments about how the world influences males and females differently, most feminist writers that I have read are basically uninterested in whether or not such differences are genetic or related to socialization. It is enough for them that the differences exist. These differences influence one's entire world and life. For those who are just thinking about this idea for the first time, I recommend that you find a little book called *The Yellow Wallpaper* (Gilman, 1973). Written about 100 years ago, it gives a picture of one woman's view of her world and, at the same time, the picture of that same world from her husband's viewpoint. Both views impress the reader with their accuracy — and they are dramatically different.

These scholars work in several areas; almost all of them are outside mathematics education. Some are trying to interpret a basic discipline of concern (such as biology or history) from a female, rather than a male, point of view. They argue that almost all scholarship, including the development of what is called science and mathematics, has been done by men and from a masculine viewpoint, utilizing values that are shared by men, but not by women. Those major bodies of knowledge that appear to be value-free and to report universal truth are in reality based on masculine values and perceptions. Since males' roles and spheres in the world have been so different from females' roles and spheres (Greene, 1984), these bodies of knowledge do not reflect 50 percent of human beings and thus are incomplete and inaccurate. Jim Schuerich (1992) has suggested that a feminist science is better than a value-free science. To support this, he draws from Charol Shakeshaft's (1987) work on educational administration and Carol Gilligan's (1982) work on the development of moral judgment. Each of them has demonstrated quite conclusively that research on male-only populations has produced results that were not only incomplete but wrong.

The idea of masculine-based interpretations in areas such as history or literature, and even in medical science, is not too difficult to illustrate nor even to accept. Many conclusions in medical research have been based solely on male subjects; their inaccuracy is easy to illustrate. History has been presented as if most of our ancestors were male and as if important things in the public arena happened predominately to males. The use of male names by female writers in order that their writing be accepted, or even published, is commonly known. Does the prevalence of this attitude apply to mathematics and if so, how? Can mathematics be seen as masculine or feminine? Is not mathematics a logical, value-free field? The idea of a masculine or a feminine mathematics is difficult to accept and to understand, even for many who have been concerned about gender and mathematics. A few people are working to explicate what a gendered mathematics might be — in particular, Suzanne Damarin (in press), Zelda Isaacson (1986), and Judith Jacobs (in press), who are struggling to define what a feminist approach to the study of mathematics education might be.

One way to approach the problem of a gendered mathematics is not to look at the subject, but to examine the way that people think and learn within the subject. This has been done in other disciplines. The work of Belenky and her colleagues (Belenky, Clinchy, Goldberger, & Tarule, 1986) in identifying women's ways of thinking and knowing has been provocative as we consider this question. Many within the field of gender and mathematics studies have interpreted this kind of basic research as necessary if we are to identify what female-friendly instruction might be — for example, the greater inclusion of co-operation rather than of competition in classrooms. Others have argued for single-sex schools oriented to the mathematics instruction of females. Running through these suggestions, it seems to me, is a basic belief that females learn differently and perform differently in mathematics than do males.

Another theme that informs many of the feminist perspectives is the necessity for females' voices to be heard (Campbell & Greenberg, 1993). To scholars with this conviction, it is not enough that researchers identify important questions, which are then studied objectively using a positivist approach. Females must have a hand in the identification of the questions; females' life experiences become critical, so that the world can be interpreted from a female perspective. So we see subjects as co-investigators, women reporting their own experiences, and women as the main subjects under investigation helping to interpret results. There are not many of these studies available currently in mathematics education, but I predict that we shall see increasing numbers of them as the importance of female voices is recognized.

It is too early to be able to assess the impact that studies using feminist methodologies will have on our understanding of the relationship between gender and mathematics, both the identification of the problem and its solutions. It appears logical to me that as I try to interpret the problem from a feminist standpoint, it is different from what I focussed on earlier. Instead of interpreting the challenges related to gender and mathematics as involving problems associated with females and mathematics, I begin to look at how a male view of mathematics has been destructive to both males and females. I begin to articulate a problem that lies in our current views of mathematics and its teaching. I am coming to believe that females have recognized that mathematics, as currently taught and learned, restricts their lives rather than enriches them.

Whatever our own value position about feminism and mathematics, I believe that we need to examine carefully how feminist perspectives can add enriched understanding to our knowledge of mathematics education. And, indeed, we should be open to the possibility that we have been so enculturated by the masculine-dominated society we live in that our belief about the neutrality of mathematics as a discipline may be wrong or, at the very least, incomplete. Perhaps we have been asking the wrong questions as we have studied gender and mathematics. Could there be a better set of questions, studied from feminist perspectives, that would help us understand gender issues in mathematics? What would a feminist mathematics look like? Is there a female way of thinking about mathematics? Would mathematics education, organized from a feminist perspective, be different from the mathematics education we currently have? Suzanne Damarin (in press) has stated that we need to "create a radical reorganization of the ways that we think about and interpret issues and studies of gender and mathematics." Many scholars believe that only as we do this will we be fully able to understand gender issues in mathematics. Perhaps my beginning to believe that the decision by females not to learn mathematics or enter mathematics-related careers because mathematics has not offered them a life they wish to lead is an indication that my old view about learning and teaching mathematics, as well as about gender and mathematics, was immature and incomplete. I am beginning to believe that an examination of what the female voices in the new research are saying will help me — and perhaps others — to understand teaching, learning, gender, and mathematics better.

NEW RESEARCH PERSPECTIVES: SOME CONTRASTS

Cognitive science and feminist perspectives, while sharing surface similarities, are based on dramatically different assumptions about females and males. These assumptions dictate the questions that are addressed, how studies are designed, and how evidence is interpreted. They are assumptions that are more far-reaching than issues of scholarship; they influence how we view the entire issue of gender and mathematics. What is the magnitude and impact of differences between females and males? Are males and females fundamentally different, so that all decisions about mathematics and understanding gender and mathematics need to be made on the basis of these differences? Or are males and females fundamentally the same, with the exception of their biological differences, and are these differences irrelevant with respect to mathematics? Cognitive science research, as it identifies universals, would suggest that looking at the world through either feminine or masculine eyes does not make sense. Feminist perspectives suggest just the opposite: female/male differences permeate the entirety of life and must be considered whenever scholarship is planned.

The implications of these assumptions are dramatic, in both doing and understanding research, as well as in work in the field of gender and mathematics. Each individual should think deeply about his or her own beliefs and reinterpret knowledge about gender and mathematics in relation to these beliefs or assumptions.

IN CONCLUSION

Research has provided rich documentation of and knowledge about variables that are related to gender and mathematics, and moderate change in what has happened to females within mathematics education has occurred partly because of this scholarship. We need to continue research that documents the status of gender differences as they exist. However, research, as we know it, must be supplemented with new types of scholarship focussed on new questions and carried out with new methodologies. Such scholarship will help in the identification of important emphases for further new research; it will also ensure that women's voices will become a major part of all educational scholarship.

While I have chosen to focus this paper on research as I perceive it, there have been other forces at work during the same time that our research knowledge has been accumulating. Many innovative interventions have been developed, based on the intuitive knowledge of concerned individuals. Mathematicians, educators, teachers, and parents have become aware of the issues related to gender and mathematics. We have come a long way. But we have a long way to go to accomplish equity in mathematics education.

NOTE

The writing of this paper was supported by the National Center for Research in Mathematical Sciences Education, Wisconsin Center for Education Research, University of Wisconsin-Madison. The Center is funded primarily by the Office of Education Research and Improvement, US Department of Education (OERI/ED) Grant no. R117G10002-93. The opinions expressed in this publication do not necessarily reflect the position or policy of OERI/ECD; no endorsement by OERI or the Department of Education should be inferred.

REFERENCES

Adetula, L. O. (1989). Solution of simple word problems by Nigerian children: Language and schooling factors. *Journal for Research in Mathematics Education, 20*, 489–497.

Anonymous. (Manuscript in preparation). Citation Classic. *Current Contents.*

Belenky, M., Clinchy, B., Goldberger, N., & Tarule, J. (1986). *Women's ways of knowing: The development of self, voice, and mind.* New York: Basic Books.

Bleier, R. (1984). *Science and gender: A critique of biology and its theories on women.* New York: Pergamon Press.

Brown, C., & Borko, H. (1992). Becoming a mathematics teacher. In D. A. Grouws (Ed.), *Handbook of research on mathematics teaching and learning: A project of the National Council of Teachers of Mathematics.* New York: Macmillan.

Campbell, P., & Greenberg, S. (1993). Equity issues in educational research methods. In S. K. Bilken & D. Pollard (Eds.), *Gender and education: Ninety-second yearbook of the National Society for the Study of Education.* Chicago: University of Chicago Press.

Carpenter, T. P., & Fennema, E. (1992). Cognitively guided instruction: Building on the knowledge of students and teachers. In W. Secada (Ed.), *Curriculum reform: The case of mathematics education in the United States.* Special issue of *International Journal of Educational Research* (pp. 457–470). Elmsford, NY: Pergamon Press.

Carpenter, T., Moser, J., & Romberg, T. (1982). *Addition and subtraction: A cognitive perspective.* Hillsdale, NJ: Lawrence Erlbaum Associates.

Casserly, P. (1980). Factors affecting female participation in advanced placement programs in mathematics, chemistry, and physics. In L. Fox, L. Brody, & T. Tobin (Eds.), *Women and the mathematical mystique.* Baltimore: John Hopkins University Press.

Damarin, S. (in press). Gender and mathematics from a feminist standpoint. In W. S. Secada, E. Fennema, & L. Byrd (Eds.), *New directions for equity in mathematics education.* New York: Cambridge University Press.

Eccles, J. S., & Blumenfeld, P. (1985). Classroom experiences and student gender: Are there differences and do they matter? In L. C. Wilkinson & C. B. Marrett (Eds.), *Gender influences in classroom interaction.* New York: Academic Press.

Fennema, E. (1974). Mathematics learning and the sexes: A review. *Journal for Research in Mathematics Education, 5*(3), 126–139.

Fennema, E., & Leder, G. (Eds.). (1990). *Mathematics and gender: Influences on teachers and students.* New York: Teachers College Press.

Fennema, E., & Peterson, P. (1985). Autonomous learning behavior: A possible explanation of gender-related differences in mathematics. In L. C. Wilkinson & C. B. Marrett (Eds.), *Gender influences in classroom interactions* (pp. 17–35). New York: Academic Press.

Fennema, E., & Peterson, P. L. (1986). Teacher-student interactions and sex-related differences in learning mathematics. *Teaching and Teacher Education, 2*(1), 19–42.

Fennema, E., Peterson, P. L., Carpenter, T. P., & Lubinski, C. A. (1990). Teachers' attributions and beliefs about girls, boys, and mathematics. *Educational Studies in Mathematics, 21*(1), 55–65.

Fennema, E., & Sherman, J. (1977). Sex-related differences in mathematics achievement, spatial visualization, and affective factors. *American Educational Research Journal, 14*(1), 51–71.

Fennema, E., & Sherman, J. (1978). Sex-related differences in mathematics achievement and related factors: A further study. *Journal for Research in Mathematics Education, 9*(3), 189–203.

Fennema, E., & Tartre, L. (1985). The use of spatial visualization in mathematics by boys and girls. *Journal of Research in Mathematics Education, 16*(3), 184–206.

Fennema, E., Wolleat, P., Becker, A., Pedro, J. D. (1980). *Multiplying options and subtracting bias.* (An intervention program composed of four videotapes and facilitators' guides.) Reston, VA: National Council of Teachers of Mathematics.

Gilligan, C. (1982). *In a different voice: Psychological theory and women's development.* Cambridge, MA: Harvard University Press.

Gilman, C. P. (1973). *The Yellow Wallpaper.* (Afterword by E. R. Hedges.) New York: Feminist Press. (Original work published 1899.)

Greene, M. (1984). The impacts of irrelevance: Women in the history of American education. In E. Fennema & M. J. Ayer (Eds.), *Women and education: Equity or equality?* Berkeley: McCutchan.

Hanna, G. (1989). Mathematics achievement of girls and boys in grade eight: Results from twenty countries. *Educational Studies in Mathematics, 20*(2), 225–232.

Harding, S. (1987). *Feminism and methodology.* Bloomington, IN: Indiana University Press.

Hopp, C. (1994). *Cooperative learning and the influence of task on learning of mathematics.* Unpublished doctoral dissertation, University of Wisconsin-Madison.

Hyde, J. S., Fennema, E., & Lamon, S. J. (1990). Gender differences in mathematics performance. *Psychological Bulletin, 107,* 139–155.

Hyde, J. S., Fennema, E., Ryan, M., & Frost, L. A. (1990). Gender differences in mathematics attitude and affect: A meta-analysis. *Psychology of Women Quarterly, 14,* 299–324.

International Commission on Mathematics Instruction. (1992). Study document for Gender and Mathematics Conference.

Isaacson, Z. (1986). Freedom and girls' education: A philosophical discussion with particular reference to mathematics. In L. Burton (Ed.), *Girls into maths can go.* London: Holt, Rinehart & Winston.

Jacobs, J. (in press). Feminist pedagogy and mathematics. *International Reviews on Mathematical Education.*

Jacobs, J. E., & Eccles, J. S. (1985). Gender differences in math ability: The impact of media reports on parents. *Educational Researcher, 14*(3), 20–25.

Koehler, M. S. (1990). Classrooms, teachers, and gender differences in mathematics. In E. Fennema & G. Leder (Eds.), *Mathematics and gender.* New York: Teachers College Press.

Leder, G. C. (1992). Mathematics and gender: Changing perspectives. In D. A. Grouws (Ed.), *Handbook of research on mathematics teaching and learning: A project of the National Council of Teachers of Mathematics.* New York: Macmillan.

Leder, G. C. (1982). Mathematics achievement and fear of success. *Journal for Research in Mathematics Education, 13*(2), 124–135.

Maccoby, E. E., & Jacklin, C. N. (1974). *The psychology of sex differences.* Stanford:

Stanford University Press.

Olivier, A., Murray, H., & Human, P. (1990). Building young children's mathematical knowledge. In G. Booker, P. Cobb, & T. N. Mendicuti (Eds.), *Proceedings of the Fourteenth International Conference for the Psychology of Mathematics Education* (Vol. 3, pp. 297–304). Mexico.

Peterson, P. L., & Fennema, E. (1985). Effective teaching, student engagement in classroom activities, and sex-related differences in learning mathematics. *American Educational Research Journal, 22*(3), 309–335.

Reyes, L. H., & Stanic, G.M.A. (1988). Race, sex, socioeconomic status and mathematics. *Journal for Research in Mathematics Education, 19*(1), 26–43.

Schuerich, J. (1992). Methodological implications of feminist and poststructuralist views of science. *The National Center for Science Teaching and Learning Monograph Series, 4.*

Secada, W. S. (1991). Degrees of bilingualism and arithmetic problem solving in Hispanic first graders. *Elementary School Journal, 52*, 306–316.

Secada, W. (1992). Race, ethnicity, social class, language, and achievement in mathematics. In D. A. Grouws (Ed.), *Handbook of research on mathematics teaching and learning.* New York: Macmillan.

Shakeshaft, C. (1987). *Women in educational administration.* Newbury Park, CA: Sage Publications.

Sherman, J. S., & Fennema, E. (1977). The study of mathematics among high school girls and boys: Related factors. *American Educational Research Journal, 14*(2), 159–168.

Stanley, J., & Benbow, C. (1980). Sex differences in mathematical ability: Fact or artifact. *Science, 210*, 1262–1264.

Walberg, H., & Haertel, G. (1992). Educational psychology's first century. *Journal of Educational Psychology, 84*(1), 6–19.

Weisbeck, L. (1992). *Teachers' thoughts about children during mathematics instruction.* Unpublished doctoral dissertation, University of Wisconsin-Madison.

MARY GRAY

GENDER AND MATHEMATICS:
MYTHOLOGY AND MISOGYNY

Gender differences in mathematics achievement — do they exist? Do they persist? Are they due to nature or nurture? Such questions have prompted a great deal of discussion and research in the 20 years since Fennema's seminal paper (1974). Regardless of the results of research, certain myths have persisted and influence the conditions for women and girls in mathematics even today.

MYTH 1: WOMEN CAN'T DO MATHEMATICS

This needs a bit of deconstruction. By mathematics, purveyors of this myth can mean anything from computation and minimally abstract algebra through geometric concepts requiring spatial perception to research at the cutting edge of the discipline. By women, even the most fervent proponent of this myth doesn't mean that no woman can do mathematics better than any man. "Women can't do mathematics" is meant primarily in the same way that "women live longer than men" was used for years to deny women equality in pension benefits (Bergmann & Gray, 1975). That is, it is meant to indicate that the "average" woman cannot do mathematics as well as the "average" man. However, something more may also be meant by the phrase "Women can't do mathematics," for it is the case that many believe that, as a matter of genetic disability, women are essentially excluded from the upper end of the mathematics achievement curve, be it represented by 800 Scholastic Aptitude Test — Mathematics (SAT-M) scores or by the tenured faculty of the country's top-rated mathematics departments.

Even though rationally this myth is understood to be a statistical statement, it has devastating effects on individuals, who interpret it to mean that *they* cannot do mathematics and as a result are either scared away or give up too easily. Differences in confidence levels between males and females still exist, to the detriment of females, even when they perform equally. These differences emerge in high school and coincide with differences in enrolment in advanced courses, differences in expressed interest in science and mathematics — which have also failed to decline over the years — and differences in access to careers in mathematics and science (Dossey et al, 1988; Horner, 1972).

However, what results do seem to show is that the cognitive differences

G. Hanna (ed.), Towards Gender Equity in Mathematics Education, 27–38.
© 1996 *Kluwer Academic Publishers. Printed in the Netherlands.*

between males and females that appear at about the time of the onset of adolescence are small, declining, and rooted in differential experiences (Friedman, 1989; Linn & Hyde, 1993). It should be noted that the differences between the scores of males and females on the SAT-Ms are idiosyncratic in several regards (Rosser, 1989). Not only has the difference between the mean scores for males and females not decreased, a difference explained only in part by the fact that more females than males take the exam, but the upper tail of the distribution is overwhelmingly male. In fact, so striking are these results and so inconsistent with other measures of mathematics achievement that their use in determining scholarship aid has been subject to legal challenges, at least one of which has been successful (*National Center for Fair & Open Testing v. Educational Testing Service and College Entrance Examination Board*, 1994; *Sharif v. New York State Board of Education*, 1988). Allegations persist that the questions on the SAT-M are biased against females; what is known for certain is that female-male differences do vary greatly from question to question (Connor & Vargyas, 1989).

Another aspect of many of the studies of gender differences in SAT-M results is that they have utilized specialized groups (largely white, middle-class and upper-middle class high-ability volunteers) so that the results — for example, that 96 percent of the 800 scores were achieved by males — may not generalize to the rest of the US population. Nonetheless, the popular press coverage of one such study has done incalculable damage to a whole generation of young women (Benbow & Stanley, 1990; Jacobs & Eccles, 1985).

Accounting for the cognitive differences that still exist is a subject of much controversy. In the Second International Mathematics Study, gender differences were not uniform from country to country, but the only discernible trait that appeared to distinguish the countries showing no gender differences from those that did was parental support for mathematics activities, a factor long thought to be crucial (Eccles & Jacobs, 1986; Hanna, 1989; Helson, 1971). At a more advanced level, the countries where the largest proportion of doctorates in mathematics go to women are those in southern Europe (30 to 40%) — where, it is conjectured, mathematics does not enjoy the status that it does in northern Europe and the US — where the proportion of women among doctorates has recently ranged from the 1 to 2 percent of Scandinavia to the 27 percent in the US. Thus the theory goes, if mathematics has low prestige, women are welcome, although an alternative explanation offered by some Italian women mathematicians is that a standardized secondary school curriculum forces women to take mathematics; doing so thus is not considered unfeminine and those who are successful continue in the field in the same numbers as do men. It should be noted, however, that even in the Mediterranean countries, the upper echelons of the mathematics establishment are male-dominated.

The paucity of women is most pronounced at the level of tenured full professors at the best institutions in the US and abroad; indeed, there are countries where no woman has attained a professorship in mathematics, albeit in systems which have proportionally far fewer at the rank of professor than in the US The proffered explanation is, of course, one of merit, tempered by the notion that until recently there have been so few women in mathematics that one could not have expected many, if any, women to have risen to these heights.

On the other hand, an alternative explanation is faith in Myth 1, underlaid as it may be with misogyny. It is difficult otherwise to understand the reason for the spate of vitriolic comments that greeted the attainment of tenure at the University of California by Jenny Harrison (Ratner, 1993). Even if one were willing to concede doubt about whether Professor Harrison deserved tenure, it would hardly have been the first instance in which a questionable tenure appointment was made. However, it is the first and indeed only case to inspire extensive Internet exchanges, letters to editors, and articles denouncing the decision.

MYTH 2: WOMEN DO MATHEMATICS DIFFERENTLY THAN MEN

Research has shown that some women respond better to non-traditional methods of teaching mathematics, but then so do some men (Belenky et al, 1986; Dumarin, 1990; Petersen & Fennema, 1985; Rosser, 1990). Other research has found no sex-based differences in, for example, problem-solving techniques (McCoy, 1994). It has been conjectured that women are more likely to try to do mathematics according to formulae, to be less willing to try different and creative techniques because of the way females are conditioned. In fact, this has been tendered as an explanation for lower SAT-M scores; speed is crucial to success on the SAT as is creative guessing, particularly working backwards from the more likely of the offered multiple-choice answers (Robinson & Katzman, 1986). At a conference on the subject of the SATs it was even proposed that SATs be untimed and "tricks" be banned as an aid to females taking them. By "tricks" the speaker meant such devices as multiplying 99 cents by 3 by subtracting 3 cents from 3 dollars (Connor & Vargyas, 1989).

Any inclination of women and girls to be more likely to follow set procedures may interact with the observed tendency of females to follow social conventions more often (Wilkinson & Marrett, 1985). It is unfortunate if males or females are taught that mathematics is nothing but the mindless application of formulae learned by rote, but there is no real evidence that females are inherently inclined to such a limited way of doing mathematics. What has been established is that students in general may have a variety of learning styles that may benefit from a variety of teaching styles. And although there is no evidence that differential treatment of girls and boys in the classroom causes differential performance, it may well lead to differential dropout rates (Fennema & Leder, 1990; Petersen & Fennema, 1985).

Moreover, women's lack of confidence may inhibit them from trying alternate approaches to problems or experimenting with different techniques (Dossey et al., 1988).

In general, there is distressingly sparse information on intervention programs designed to increase the participation of women in mathematics; most have not been accompanied by long-term studies of effects. Very few programs use the random assignment necessary for reliable assessment; there is little cross-institutional research and hardly any longitudinal studies, even though the choice to do mathematics is made over time. There is also a critical need for carefully designed observational studies to assess in depth the impact of such things as classroom dynamics.

MYTH 3: WOMEN DO DIFFERENT MATHEMATICS THAN MEN

Feminist critiques of mathematics, or more generally of science, often contend that women's differential success rate is due to the subject matter, not to the women (Gilligan, 1982); the corollary is that mathematics, not women, must change (Keller, 1985). In fact, Sandra Harding (1991) has said that "women scientists" is a contradiction in terms and would presumably characterize "women mathematicians" similarly. Unfortunately, most of the critics have so little understanding of mathematics that their theory is divorced from the reality of the subject.

A more traditional version of the myth that women do different mathematics is that their work is concentrated in a few areas somehow peculiarly suited for the female mind. An examination of the field of specialties of women some 15 years ago did show, at least within the US, a disproportionate concentration of women PhDs in algebra. Some theorized that this reflected the influence of Emmy Noether, the "father of modern algebra" and the best known woman mathematician, as a role model. However, closer examination and conversations with the women themselves revealed that many of the women had the same handful of dissertation advisers. Another example of concentration in a single field was largely due to 16 women students of the same person; when the universe is so small, such artifacts occur.

For a number of years, the yearly production of women PhDs in mathematics hovered around 85, with a spurt in 1993 to 124 (AMS-IMS-MAA, 1993), but more recently their distribution by field has not differed significantly from that of men. As the most well-documented sex-related differences in mathematics concern spatial perception, one might expect to find few women in geometry and topology, but this is not the case. Current thinking is that the differences in spatial perception may be related to differential participation in sports, a phenomenon becoming less pronounced (Fennema & Sherman, 1978; Linn & Hyde, 1993; Linn & Petersen, 1985; Sherman & Fennema, 1977).

What has been observed is that women may be less inclined to do their doctoral work in a "hot" field, due possibly to either greater independence or differential access to mentors, and that this may hinder their chances to get a job. The single most important obstacle to real affirmative action is that mathematics departments generally seek to employ someone in a very narrowly defined field; given the general scarcity of women, this impacts heavily on their chances (Uhlenbeck, 1993).

MYTH 4: MATHEMATICIANS DO THEIR BEST WORK WHEN THEY ARE UNDER 40

This might seem unrelated to gender, but in fact, anecdotal evidence shows that whereas male mathematicians report getting their best results between the ages of 25 and 40, females generally find that their best research occurs between 35 and 50 (American Mathematical Society, 1991; Stern, 1978). There are several possible explanations — that women get started later, that they take time out for child-bearing and child-rearing, and that they continue to work as active researchers at an age when men have diverted their efforts to administration of various sorts. That discrimination may preclude women's taking on administrative roles may be good for their mathematics!

Because of the belief that those over 40 are past their prime, women whose research peaks at a later date may never receive the recognition they deserve. For example, the Fields medal, mathematics equivalent of the Nobel prize, is given exclusively for work done before a mathematician reaches the age of 40. More prosaically, women who receive their PhDs at a later age than usual are often at a disadvantage in the job market.

MYTH 5: MATH IS TOUGH!

It seems only fair that at least one myth ought to be a reality, and this one is. There was much criticism several years ago when a new Barbie doll came equipped with the oral pronouncement: "Math class is tough!" (Leggett, 1992) Whereas this was clearly sexist — Ken was not programmed with such a saying — and unnecessarily discouraging to girls, it is not untrue. For many girls — and boys — math class is tough, and to suggest otherwise just makes it all the more likely that girls will become discouraged and drop out if they are led to believe that it should be easy. And society does make it easier for girls to give up, particularly in a field where they are thought not to be any good anyway.

One cure for "math anxiety" at the time it was thought to be endemic in the female population was for women to tell themselves that they were no longer anxious (and to sit in white instead of blue chairs); unfortunately that did not improve their mathematics skills (Tobias, 1978). What mathematics requires for most people is concentrated, persistent effort; to imply otherwise does no one a favor. On top of any inherent difficulty, there is a constant battle to overcome the effects of prejudice and to balance professional and personal demands. One of the two women mathematicians currently in the US National Academy of Science put it well: "I cannot think of a woman mathematician for whom life has been easy. Heroic efforts tend to be the norm" (Uhlenbeck, 1993). Things have not changed much since 1888. Contestants for that year's Prix Bordin of the French Academy submitted their papers anonymously, but included an identifying phrase. Sonya Kovalevsky (also known as Sofia Kovalevskaia), an outstanding mathematician who was never able to get a job in her native Russia, wrote on her prize-winning entry, "Say what you know, do what you must, come what may."

What "math is tough" does not mean, however, is that it is not for women. Of course, it used to be thought that any intellectual endeavor was too much for women — to find math problems in the *Ladies' Diary* published from 1704 to 1841 comes as a real shock (Perl, 1979). The belief that math is too hard for women may be just another manifestation of the misogyny rampant in the field of mathematics. A colloquy between two distinguished mathematicians, at the time when the establishment of a Committee on Women of the American Mathematical Society was being contemplated, went as follows: Cathleen Morawetz of NYU, "You may not have noticed that there are not many women mathematicians." Saunders MacLane of Chicago: "Well, mathematics is a very difficult subject." (American Mathematical Society, 1991) Some 20 years later MacLane defended his department's lack of women faculty by remarking that they had once (40 years earlier) hired a woman at Chicago but her research wasn't very good. Some things never change.

MYTH 6: THE DAMAGE IS DONE BEFORE WOMEN EVER GET TO COLLEGE

It used to be that a majority of girls failed to take more than two years of secondary school mathematics so that advanced algebra became a "critical filter" that prevented them going on to study mathematics or science in college (Chipman, Brush, & Wilson, 1985). This is no longer the case; the substantial majority of college-bound females and males have three years of high school mathematics. What is different is that women more frequently come to college without a definite major in mind so that a discouraging experience in a mathematics class may be more likely to have an impact on their future plans. Thus, what happens to women as undergraduates is important. Women represent nearly 50 percent of those getting a bachelor's degree in mathematics in the US, but few of them go on in mathematics. The largest single differential dropout rate between the sexes occurs between the baccalaureate and the doctorate. Ever since 1983, in the US, the percentage of mathematics doctorates that are earned by women has been over 20, having gradually risen from the 6 percent prevalent from 1930 through the mid 1960s. (In fact, the *number* of women getting PhDs in mathematics has remained relatively constant for many years; the change in percentage is due to the decrease in the number of men.) Thus, were efforts to be concentrated where women are most affected, it would be on doctoral programs; graduate faculty bear a heavy responsibility for increasing the proportion of women — and non-Asian minorities — among mathematicians.

A great deal can be done — and in fact, many institutions are doing it — as the percentage of women among mathematics PhDs varies greatly from university to university (American Mathematical Society, 1991). Obviously financial assistance is important, as is a supportive atmosphere. Sometimes women have trouble establishing the close relationship mentoring requires, and in fact with the attention on harassment, male faculty report shrinking from working closely with women for fear of misperceptions that might arise.

Seeking students from the liberal arts colleges, where good women students are more likely to go as undergraduates, is important. Moreover, particularly if there is a commitment to racial as well as gender diversity, recruitment efforts need to focus on those who have been out of school for awhile. If women students come with less extensive or less recent undergraduate preparation, they need to be placed in proper prerequisites and given time to acquire the necessary background. That mathematics is not necessarily a solitary pursuit may need to be demonstrated to some to whom such a career would not appeal; solving real-world problems and helping others may appeal to some whereas the pleasure derived from an elegant proof may be the major attraction for others. Demonstrating the variety of experiences and pleasures that one may derive from mathematics draws not only women but also more men into mathematics.

One of the most acute needs for research is an in-depth longitudinal study of women in graduate school in mathematics and the factors that contribute to their retention. Although it is known that a substantial number of women faculty is not necessary in order for a supportive atmosphere to exist and that particular faculty have had particular success in mentoring women students, very little else is known.

When the Association for Women in Mathematics was founded some 24 years ago, its manifesto, thought at the time to be exceedingly radical included:

1. Equal consideration for admission to graduate school
    and support while there.
2. Equal consideration for faculty appointments at all levels.
3. Equal pay for equal work.
4. Equal consideration in assignment of duties, for promotion,
    and for tenure.

5. Equal consideration for administrative appointments at all levels in universities, industry, and government.
6. Equal consideration for government grants, positions on review and advisory panels, and positions in professional organizations.

The last five will be considered later, but not all that much progress has been made even on the first. Whereas women graduate students are no longer absolutely denied financial aid, they are disproportionately given teaching assistantships rather than research fellowships, by both universities and the National Science Foundation. For example, in a recent competition, the success rate for men applicants for NSF graduate fellowships in mathematics was three times as great as for women, a result unique to mathematics. In other fields, the success rate was gender-neutral (except in chemistry, but even there it was better than in mathematics). Asked to explain women's lack of success in mathematics, those on selection panels pointed most frequently to the differences in letters of reference, clearly a source of subjectivity or even discrimination.

MYTH 7: WOMEN ARE GETTING ALL THE JOBS

With the tightening of the job market has come the apparently inevitable backlash. No doubt there are young white men who have been told that they cannot be hired because of affirmative action requirements, and in fact there may be cases where that was true (Jackson, 1993). However, as the percentage of 1993 women PhDs who had positions as of fall 1993 was 88.2 percent, compared with 85.5 percent for men, it cannot have happened very often. More likely, as has been the case for the past 25 years of affirmative-action obligations on departments, men were told that they were turned down for affirmative action reasons as a gentle let-down or even to create discontent. In fact, this year was the first time that the top departments hired women at the rate that they produced them — a levelling of the playing field long sought, but hardly a tilt towards women. To illustrate how glacial progress in some quarters has been: In 1928 20 percent of the University of California, Berkeley, mathematics faculty were women; in 1948 it was 7 percent, and in 1968 it was 0 percent. Since then, the percentage has not risen to the level of 1948, much less to that of 1928.

It certainly is the case that women did not get all the jobs in the past, especially not all the best jobs. In spite of the previously cited figures on PhD production, only 5 percent of the tenured faculty at doctorate-granting institutions are women. In the top ten mathematics departments, barely 1 percent of the tenured faculty are women and the situation for untenured tenure-track faculty is not much better, even though 12 percent of the PhDs from these departments went to women in the decade of the 1980s. Women are disproportionately concentrated at bachelors- and masters-granting

institutions, where teaching loads are heavier, support for research generally less, and salaries lower. Even when women are at doctorate-granting
institutions, they frequently have heavier teaching loads, lower-level courses, and less access to doctoral students. So much for item two of the manifesto mentioned above.

In 1993, starting salaries for women and men PhDs in mathematics were
relatively close — women getting $200 less on average (AMS-IMS-MAA,
1993). However, the gap widens with experience (Ahern & Scott, 1981). In
general, women full professors make 10 percent less than their male counterparts (American Mathematical Society, 1991). A possible explanation is
the fact that women publish less; however, whether that is due to lesser
abilities or to job conditions or to outright discrimination is not clear. For
example, when the same paper was submitted with names of men and
women attached, reviewers rated the version with a man's name higher than
the version identified as being by a woman (by both men and women
reviewers, although the sex-difference was smaller in the case of women
reviewers) (Paludi & Bauer, 1983).

Another factor in getting and keeping good jobs is the dissemination and
recognition of one's research. In the past three years, 16 percent of the
speakers in special sessions of meeting of the American Mathematical
Society were women in those sessions organized or co-organized by a
woman; in those organized by men the percentage was only half as great
and in one-third of the sessions there were no women speakers. Whether
these figures resulted from inadvertence or discrimination, in an effort to
provide for more equity, the governing body of the American Mathematical
Society passed a resolution requiring session organizers to seek out women
as speakers. In fact, just recently a session organizer was forced to resign
from the task when he announced his unwillingness to comply with the
mandate.

Both the dearth of women speakers and the failure to hire more women
can be at least in part attributed to a widely observed phenomenon —
namely, that institutions or session organizers seek only the top handful of
women mathematicians; if they are unavailable, no further efforts are made,
although if the first choice men cannot be obtained, others are found. For
example, doctoral institutions are divided into three categories (in declining
order of prestige); those in all three categories tend to hire women PhDs
only from category I, whereas category II and III institutions hire men from
their own level. Notably, 1993 was the first year in which the percentage of
women among new PhDs hired by category I institutions equalled or
exceeded the percentage of women among new PhDs produced by category
I institutions.

Mathematics departments have acquired a reputation for being particularly recalcitrant about engaging in affirmative action (Selvin, 1992; Uhlenbeck, 1993). It is tempting for those responsible to believe that this is pure-
ly a question of quality — of clearly defined and rigidly adhered to stan-

dards, and not one of hiring someone who seems to belong, who fits their image of a mathematician. However, hiring judgments are not infallible — if they were, everyone would get tenure! Neither are decisions as to whether to make a particular grant or to publish a specific paper. Subjectivity clearly enters into each of these. Can the irrefutable evidence that over the years these decisions have had a cumulative adverse impact on women be attributed to the fact that women are inferior, to inadvertence, or to decision-makers' (be they men or women, for women can also be misogynists) belief that women are inferior?

The attempt to find what really motivates hiring, tenure, and similar decisions, to discover whether those attracted to mathematics are inherently inclined to hold views that have led to making the profession look as it does, or whether they acquire them with the ability to prove theorems, would be fascinating subjects for research. And whatever is found out, is the situation immutable? An interesting sidelight, perhaps worthy of exploration, is the fact that together with the physical sciences, the discipline of philosophy most resembles mathematics today from the point of view of the sex and race composition of its practitioners.

CONCLUSION

The disadvantages faced by women in mathematics in career access and earning power are not accounted for by cognitive and psycho-social gender differences (Linn & Hyde, 1993). Thus the mythology does not explain the reality. Does the misogyny?

REFERENCES

Ahern, N. C., & Scott, E. L. (1981). *Career outcomes in a matched sample of men and women Ph.D.'s: An analytical report*. Washington, DC: National Academy Press.

American Mathematical Society. (1991). Special issue: Women in mathematics. *Notices of the American Mathematical Society, 38*(7), 701–754.

AMS-IMS-MAA Data Committee. (1993). 1993 Annual AMS-IMS-MAA Survey. *Notices of the American Mathematical Society, 40*(9), 1164–1172.

Belenky, M. F., Clinchy, B. M., Goldberger, N. R., & Tarule, J. M. (1986). *Women's ways of knowing: The development of self, voice and mind*. New York: Basic Books.

Benbow, C., & Stanley, J. C. (1980). Sex differences in mathematical ability: Fact or artifact?" *Science, 210*, 1262–1264.

Bergmann, B. B., and Gray, M. W. (1975). Equality in retirement benefits. *Civil Rights Digest, 8*, 25–27.

Chipman, S. F., Brush, L. R., & Wilson, D. M. (Eds.). (1985). *Women and mathematics: Balancing the equation*. Hillsdale, NJ: Lawrence Erlbaum Associates.

Connor, K., & Vargyas, E. J. (1989). *Gender bias in standardized testing.* The Proceedings of a Hearing co-sponsored by the National Commission on Testing and Public Policy and the National Women's Law Center. Washington, DC: National Commission on Testing and Public Policy.

Dossey, J. A., Mullis, I.V.S., Lindquist, M. M., & Chambers, D. L. (1988). *The mathematics report card. Are we measuring up?* Princeton: Educational Testing Service.

Dumarin, S. K. (1990). Teaching mathematics: A feminist prospective. In T. J. Cooney (Ed.), *Teaching and learning mathematics in the 1990s: 1990 Yearbook* (pp. 144–151). Reston, VA: National Council of Teachers of Mathematics.

Eccles, J. S., & Jacobs, J. E. (1986). Social forces shape math attitudes and performance. *Signs, 11,* 367–389.

Fennema, E. (1974). Mathematics learning and the sexes: A review. *Journal for Research in Mathematics Education, 5,* 126–139.

Fennema, E, & Leder, G. C. (Eds.). (1990). *Mathematics and gender.* New York: Teachers College Press.

Fennema, E., & Sherman, J. A. (1978). Sex-related differences in mathematics achievement, spatial visualization, and sociocultural factors. *American Educational Research Journal, 14,* 51–71.

Friedman, K. (1989). Mathematics and the gender gap: A meta-analysis of recent studies of sex differences in mathematical tasks. *Review of Educational Research, 59,* 185–213.

Gilligan, C. (1982). *In a different voice: Psychological theory and women's development.* Cambridge, MA: Harvard University Press.

Hanna, G. (1989). Mathematics achievement of girls and boys in grade eight: Results from twenty countries. *Educational Studies in Mathematics, 20*(2), 225–232.

Harding, S. (1991). *Whose science? Whose knowledge?* Ithaca, NY: Cornell University Press.

Helson, R. (1971). Women mathematicians and the creative personality. *Journal of Consulting and Clinical Psychology, 36,* 210–220.

Horner, M. S. (1972). Toward an understanding of achievement-related conflicts in women. *Journal of Social Issues, 28,* 157–175.

Jackson, A. Are women getting all the jobs? *Notices of the American Mathematical Society, 41*(4), 286–287.

Jacobs, J. E., & Eccles, J. S. (1985). Gender differences in math ability: The impact of media reports on parents. *Educational Researcher, 14*(3), 20–25.

Keller, E. F. (1985). *Reflections on gender and science.* New Haven: Yale University Press.

Leggett, A. (1992). Barbie. *AWM Newsletter, 22*(6), 12.

Linn, M. C., & Hyde, J. S. (1993). Gender, mathematics, and science. *AWM Newsletter, 23*(3), 17–23; *23*(4), 14–17.

Linn, M. C., & Petersen, A. C. (1985). Emergence and characterization of gender differences in spatial ability: A meta-analysis. *Child Development, 56,* 1479–1498.

McCoy, L. P. (1994). Mathematical problem-solving processes of elementary male and female students. *School Science and Mathematics, 94*(5), 266–270.

National Center for Fair & Open Testing v. Educational Testing Service and College Entrace Examination Board (D.D.C. April 29, 1994).

Pauldi, M. A., & Bauer, W. D. (1983). Goldberg revisited: What's in an author's name? *Sex Roles, 9*, 387–390.

Perl, T. (1979). The Ladies' Diary or Women's Almanack 1704–1841. *Historica Mathematica, 6*(1), 36–53.

Peterson, P. L., & Fennema, E. (1985). Effective teaching, student engagement in classroom activities, and sex-related differences in learning mathematics. *American Educational Research Journal, 22*, 309–336.

Ratner, M. (1993). Letter to the editor. *AWM Newsletter, 23*(5), 17.

Robinson, A., & Katzman, J. (1986). *Cracking the system.* New York: Villard.

Rosser, P. (1989). *The SAT Gender Gap: Identifying the causes.* Washington, DC: Center for Women Policy Studies.

Rosser, S. V. (1990). *Female-friendly science and applying women's studies methods and theories to attract students.* New York: Pergamon Press.

Selvin, Paul. (1992). Heroism is still the norm. *Science, 255*, 1382–1383.

Sharif v. New York State Board of Education, 88 Civ. 8435 (JW) (S.D.N.Y. 1988).

Sherman, J., & Fennema, E. (1977). Sex-related differences in spatial visualization, and affective factors. *American Educational Research Journal, 14*, 51–77.

Stern, N. (1978). Age and achievement in mathematics: A case study in the sociology of science. *Social Studies of Science, 8*, 127–140.

Tobias, S. (1978). *Overcoming math anxiety.* New York: Norton.

Uhlenbeck, K. (1993). Address to Department Chairs' Colloquium of Joint Policy Board of Mathematical Sciences. Arlington, VA.

Wilkinson, L., & Marrett, C. (Eds.). (1985). *Gender influences in classroom interaction.* Orlando, FL: Academic Press.

GILAH C. LEDER

GENDER EQUITY: A REAPPRAISAL

Equity: Fairness; recourse to principles of justice to correct or supplement law; system of law coexisting with and superseding common and statute law.

                                                    −*Oxford English Dictionary*

In many countries, there are no longer any formal barriers to education or careers based on sex, class, race, or religious beliefs. Educationally, females are not a typically disadvantaged group. They often do well at school and in many countries now have higher retention rates than males. Females are certainly capable of doing mathematics, although they are somewhat more likely than males to opt out of the most demanding mathematics courses, both in school and beyond, and to participate less frequently than males in educational and career areas for which strong mathematical qualifications are a prerequisite.

Mathematics, gender, and equity issues have attracted much research attention in recent years and are likely to continue to do so. It is a far cry, conceptually, from sex differences in mathematics learning (a popular topic two decades ago) to our current considerations of equity and diversity with respect to gender. The variety of paradigms, research techniques, and settings in which relevant research has been carried out has heightened our appreciation of the myriad of direct and indirect factors that may help explain, for example, the subtle differences in males' and females' attitudes to mathematics which continue to be reported in the literature or the persistent gender differences in participation in the most demanding and advanced mathematics courses, already referred to above.

In early work, there was much emphasis on recording gender differences in mathematics performance and participation in advanced mathematics courses. Likely explanations for the differences observed and the effectiveness of various intervention strategies soon began to be documented. That males' experiences, behaviors, and beliefs should be accepted as the norm and that females should be provided with equal opportunities to attain the standards defined by males was rarely questioned. The removal of barriers and — if necessary — the resocialization of females were seen as paths to equity.

With the growth, in volume, sophistication, and stature, of feminist research came increased acceptance that females' values and experiences, when different from those of males, should be considered of equal worth. Equity for females, it is now more commonly argued, may require a change of current social structures and a re-evaluation of popular value positions

G. Hanna (ed.), Towards Gender Equity in Mathematics Education, 39–47.
© 1996 *Kluwer Academic Publishers. Printed in the Netherlands.*

and norms. This debate allows the assumptions that underpin a range of initiatives aimed at achieving equity in mathematics to be re-examined. In this chapter, some common definitions of feminism found in the research literature are linked to the portrayal of gender issues in the popular press and compared with the format of different mathematics programs familiar to educators, parents, and students. Media reports are chosen as the linking vehicle because of their role in reporting the findings of research to their readers in simplified terms, their considerable penetration, and their capacity to mould and reinforce popular opinion, values, and expectations.

GENDER DIFFERENCES IN MATHEMATICS LEARNING: ALTERNATE PARADIGMS

It is convenient to consider, as a starting point, the four theoretical perspectives on gender differences in mathematics used in the International Organisation of Women and Mathematics Education (IOWME) Study Group (one of the strands organized as part of the 7th International Congress on Mathematical Education [ICME–7] held in Quebec in 1992): the intervention perspective, the segregation perspective, the discipline perspective, and the feminist perspective. At the time, Mura (1992) pointed to the inadequacy of these groupings for describing the theoretical models and initiatives to which they referred. The categories showed inevitable overlap. In particular the feminist perspective included a number of approaches based on conflicting assumptions, many of which could have been included in one of the other sections. Mura asked:

> Why, I wondered, had only one of these four perspectives been labelled feminist? And why that particular one? Isn't the issue itself of "gender imbalance" a feminist issue? And won't all approaches seeking to redress the imbalance correspond to some kind of feminist analysis and practice? (p. 1)

Conversely, she could have asked: Why does the category "feminist perspectives" contain references to research and practices that are based on conflicting assumption about gender, gender imbalances, and the ideals to which we should aspire?

Many have set about the task of defining feminism in its various forms. Kristeva (1981) distinguished between "the first wave or egalitarian feminists demanding equal rights with men ... the second generation, emerging after 1968, which emphasised women's radical difference from men ... [and the] new generation of feminists" (p. 187) which will have to reconcile the two previous approaches. The notion of first, second, and third generation feminists has also been adopted by Noddings:

> In the first generation, women seek equality with men; this is the typical liberal position. In the second, they ... reject uncritical assimilation into the male world.... In the third, women critique what they sought and accomplished in the first two phases and seek solutions that arise out of a careful synthesis of old and new questions. (1990, p. 393)

Alternate labels — those of "liberal," "social," and "radical" feminisms — have been used by Kenway and Modra (1992). For these researchers, liberal feminism is consistent with Kristeva's first generation and radical feminism with her second generation feminisms, while social feminism refers to an approach with more Marxist roots, aiming to "[end] education's complicity in reproducing the complex, intersecting social relationships which are class and gender" (Kenway & Modra, 1992, p. 139). Mura's (1992) clusters of feminism of equality, radical feminism, and feminism of difference can be aligned, respectively, with Kenway and Modra's liberal, social, and radical feminisms. And Offen (1988), in an historical overview that highlights the confusion of species in this taxonomy, based her distinction on different understandings of the basic unit of society — either the family or the individual — to come up with relational feminisms or individual feminisms. The first of these stresses the difference between males and females and the way the two groups complement each other; the second celebrates

> the quest for personal independence in all aspects of life, while downplaying, depreciating, or dismissing as insignificant all socially defined roles and minimising discussion of sex–linked qualities or contributions. (p.136)

Offen's distinctions are consistent with Kenway and Modra's radical and liberal feminisms and with Mura's feminism of difference and feminism of equality, respectively.

All models are reductive and, therefore, open to criticism. The purpose of labels and categories is to facilitate ease of understanding and communication. The aim of the next section is to highlight the complexities that often become invisible if the above and similar models are not problematized, and to propose a not so neat but more workable distinction between theoretical approaches to gender and the practices they support.

THEORY INTO PRACTICE

A recent survey of the popular press (September 1994) in one Australian capital city examined more than 1000 articles on women and gender issues. Examples taken from these data are used as a reflection of the range of opinions and approaches to gender equity issues that can be found in Australia today. The most striking characteristic of this survey was the sheer number of articles identified — over 1000 from three newspapers over a four week period — indicating just how frequently women's issues are referenced in the media.

The Print Media: Impact

Concern is often expressed about the way in which issues are reported by the media. Windschuttle (1985), for example, argued that "when we compare news reporting in a number of subject areas with informed opinion originating outside the media there is often very little connection" (p. 262). The effects of subjective accounts are widespread. A report on discrimination and religious conviction concluded:

> Many feature articles about minority religious groups re–interpret new information to conform with old material held on file, whose tone is usually pejorative. Negative stereotypes are perpetuated, the more so as what is already in print tends to be believed. The same material is also repeated and distorted (New South Wales Anti–Discrimination Board, 1984, p. 195).

The implications of biased reporting on women's lives have not gone unnoticed:

> The Victorian era gave rise to mass media and mass marketing — two institutions that have since proved more effective devices for constraining women's aspirations than coercive laws and punishments. They rule with the club of conformity, not censure, and claim to speak for female public opinion, not powerful male interests. (Faludi, 1991, p. 48)

Faludi further argued that incalculable harm was being done by contemporary media stories:

> The press was the first to set forth and solve for a *mainstream* audience the paradox in women's lives ... women have achieved so much yet feel so dissatisfied; it must be feminism's achievements, not society's resistance to these partial achievements, that is causing women all this pain (p. 77, italics added).

Further acceptance of the power of the media can be inferred from the well-funded advertising campaign "Maths multiplies your choices" devised by an Australian State government (Department of Labour, 1989) to attract more females into mathematics courses and related areas. Schools reported an increased awareness among parents of the role played by mathematics as a critical filter to further educational and careers options. Greater retention of females in post-compulsory mathematics courses was reflected in subject enrolment figures. However, initiatives taken to increase females' participation in mathematics can be overtaken by events which occur in the wider context. As youth unemployment has increased in industrialized countries, many disillusioned youth ignore arguments which stress that studying mathematics will lead to increased employment opportunities. In this climate, previously successful strategies which focussed on the usefulness of mathematics become irrelevant.

The educational role of the media and its impact on aspects of mathematics learning can also be inferred from the long-running program "Sesame Street." The show has been accredited with "a special status as part of the cultural milieu from which children are thought to acquire a wide range of early childhood competencies — early number competencies among them" (Secada, 1992, p. 641).

The Print Media: Content

The articles collected with a focus on gender issues were grouped according to their major theme. A diversity of opinions and approaches was observed, as can be seen from the categories listed below.

Resocialization

Included here are those who advocate, for the betterment of women and society, the promotion in females as well as males of attributes traditionally associated with males. This resocialization of girls and women is an element of Mura's feminism of equality and Kenway and Modra's liberal feminism. In mathematics education, this strategy would support the development of competitiveness and risk-taking in girls as a means of achieving their more active participation in class. Newspaper examples of the assumption that masculine traits should be encouraged and valued in females include

> ... the three women participants who spoke exhibited true grit — determination, obstinance and a willingness to persevere. (*The Australian*, September 14, 1994 p. 31)

> In our applications laboratory in our first semester women find they are just as good at mechanical work and they enter the following semester with a confidence equal to that of the men. (*The Australian*, September 3, 1994, p. 20)

This approach is open to criticism because it privileges so-called masculine characteristics and fails to acknowledge the value of feminine traits. It is also at odds with those who hypothesize that females and males have different preferred learning styles which should be recognized in the way mathematics is presented in the classroom.

Removal of Barriers

Grouped in this category are advocates of anti-discrimination and equal opportunity legislation, affirmative action, and the removal of barriers identified as inhibiting females' participation in those areas traditionally dominated by males. These measures are also consistent with the aims of feminism of equality, liberal feminism or, indeed, with an intervention perspective. These approaches are based on liberal notions of individual rights and competition. They assume that, if these rights are respected, individuals will be free to achieve to their full potential. Differences in outcomes are expected and seen as the healthy result of differences in natural potential and personal preference, but there should be no significant and consistent differences between different societal groups. Newspaper examples of this approach include

> At its forthcoming annual conference, the Labor Party will no doubt engage in a messy civil war over affirmative action proposals aimed at getting more women in to politics. (*The Australian*, September 3, 1994, Review, p. 1)

> ... the proportion of women in our parliament remains low — obviously, there are still formidable barriers to their political advancement. More women should be involved in politics across the spectrum. If setting quotas can remove such barriers, it will be an advance for our democracy. (*The Herald Sun*, September 3, 1994, p. 2)

Initiatives specifically aimed at increasing girls' participation in existing mathematics courses are consistent with this approach. Moves to eliminate sexism from mathematics texts and tests, to stop sexual harassment inside and outside the mathematics classroom, and re-timetabling of classes to facilitate females' selection of mathematics classes are examples of such strategies.

Difference Approach

The themes in this category are described by Kenway and Modra (1992) as radical feminism, by Mura (1992) as feminism of difference, and overlap with Offen's (1988) conception of relational feminism. Recognition of the conflicting subcategories — called here *social difference, essential difference*, and *anti-feminist* — highlight the inadequacy of a model that does not discriminate clearly between them.

Social difference: Those advocating — whether in the educational literature or the print media — the revaluing and promotion of attributes traditionally associated with females (e.g., co-operation, nurturing) in both males and females for the betterment of society fall under this heading. Print media articles with this theme include:

> A study that will attempt to find out why women are not attracted to engineering may result in making the subject more appealing for both sexes.... The idea that we have to change women, that we have to get women to change their ideas about whether they can cope with engineering, is a bit passé. (*The Australian*, September 9, 1994, p. 27)

> ... women aren't perceived as powerful or as having a network and boards are looking for powerful people.... A problem with boards ... is that they are often not strategically selected. Creative people, the people with marketing skills or people with management skills are left out. (*The Australian*, September 17, 1994, p. 3)

In the mathematics classroom, this approach is reflected in programs which encourage co-operation, group work, and less competitive activities in co–educational classes. Recent changes to large-scale external examinations to include project work and investigative problem-solving tasks to be done over an extended period of time as well as the more traditional, timed tasks (Leder, 1994) are another example.

Those who adhere to the social difference view criticize the use of single–sex classes as a means of redressing gender imbalances.

> The single–sex strategy in its most popular manifestation is unlikely to change the educational opportunities of girls in any fundamental way because it focuses exclusively on changing the behaviour of girls. It neglects the much more difficult problem of changing the attitudes and behaviours of boys and teachers, and the nature of the curriculum itself. (Willis & Kenway, 1986, pp. 137–138)

Essential difference: Implicit in the articles and strategies which fall in this category is the assumption that males and females have innate, different qualities, skills, and approaches. While all such attributes should be valued, not all will be desired or considered attainable by both sexes.

Both the social and essential difference approaches are consistent with Mura's (1992) feminism of difference and Kenway and Modra's (1992) radical feminism, and with Offen's (1988) relational feminism. Although the recognition and valuing of females' contributions, and their ability to enrich and transform traditionally male-dominated areas, is common to both difference approaches, the second lends itself to a segregation perspective — that is, a more separatist approach. In mathematics education, actions consistent with the essential difference stance support single-sex classes as well as a reconsideration of the scope and content of the mathematics curriculum. For example, Willis (1989) maintained that a curriculum embedded

> in real-world social concerns and in people-oriented contexts; presenting it as making "humansense", non-arbitrary, non-absolute and also fallible; and presenting a social-historical perspective to help students become more aware of the "person-made" quality of mathematics. (p. 38)

can be achieved and is more likely to be inclusive of girls. However, the possibility that such a course might still be dubbed a "soft" version of the mainstream must also be recognized. Comparisons of the advantages and disadvantages of different educational contexts and settings should acknowledge the influence of many complex interacting and often subtle factors.

Anti–feminist: Writings which promote the return to traditional gender roles and argue that feminist agendas result in the deterioration of social values to the detriment of men and boys are referred to in this category. This approach overlaps with Offen's (1988) relational category, since it assumes that males and females are different and should fulfil different but complementary roles. A typical plea from those advocating this approach is a return to the marriage between woman as the homemaker and man as the provider. This position does not sit well with examples above that similarly stress the differences between men and women, yet have more liberal aims. Promotion of these themes continue:

> In the brave new world of so called "sexual equality", men have become superfluous to women, society and, worse still, to their own families. Unless the drive to equalise the roles of the sexes is stopped, family life will continue to decline. (*The Herald Sun*, September 3, 1994 p. 94)

> A pretty secretary who flaunts her attributes may get promoted. An ugly male who leers at those attributes will be charged with sexual harassment. When a middle-aged woman feels trapped halfway up the corporate ladder, it is called a glass ceiling. When a middle-aged man feels trapped halfway up the corporate ladder, he is called a failure. According to Naomi Wolfe and Kaz Cooke, it is a patriarchal plot that forces women to dress up, wear make-up and look good. It is not, however, a matriarchal plot that forces men to wear ties, suits, overalls and die seven years earlier. (*The Herald Sun*, September 4, 1994, p. 30)

Socialist Approach

Reflected here are the views of those who interpret the dominant institutions and organizations as functioning to perpetuate gender imbalances in order to maintain their own power through the consented control of women and other groups via exploitation, oppression, and various modes of enculturation, such as religion, education, and fashion. This Marxist interpretation is consistent with Mura's (1992) radical feminism and Kenway and Modra's (1992) social feminisms, and, as with a discipline perspective, leads to a questioning of the very nature of school mathematics, its content, teaching, and assessment.

Examples of such writings from the newspaper sample include:

> In societies where the majority of men have little power or wealth and where they live in fear of oppressive governments or powerful neighbours, the only way they find to assert their manhood, their honour, is by controlling women. (*The Australian*, September 3, 1994, Review, p. 6)

> No matter how good we women get at earning our own living and looking after ourselves, we are primed from infancy to feel less than complete without a male partner — and to know that walking around with your backside slapping against the back of your knees can put all that in jeopardy. We don't just see ourselves in the mirror, we invariably see ourselves through the eyes of men. (*The Herald Sun*, September 1, 1994, p. 15)

The survey also revealed that stereotyped language continues to be used:

> ... Chairman of the Victorian Bar Council Mrs Sue Crennan, QC ... (The Herald Sun, September 13, 1994)

> Dr Ruth Shatford, the commission's chairwomen, said there was too much emphasis in the present prayer book about God being linked to images of power, maleness and hierarchy. She said such images alienated many women. (*The Age*, September 3, 1994, p. 5)

FINAL COMMENTS

The theoretical categories used in this chapter are not mutually exclusive. They do not describe completely the perspectives of particular individuals or groups or the assumptions behind particular actions or initiatives. These are always more complex. Many of the print media reports combine elements of two or more approaches.

As well as highlighting the need to be critical about the strengths, weaknesses, and assumptions found in any one theoretical approach, the sample of articles discussed reflects the mixed and often conflicting messages offered by the print media about the position of females in society. Similar mixed messages are conveyed by the different strategies collectively advocated and implemented to achieve equity in mathematics learning. Mapping perspectives of feminisms, and their portrayal in the print media, onto educational initiatives to increase equity in mathematics education has made more explicit the theoretical assumptions and expectations of the different programs as well as the reasons for their varied success to date. The exercise, it is hoped, will also facilitate debate between those working within different paradigms yet with a common goal of achieving equity in mathematics education.

NOTE

The financial support of the Australian Research Council and the contribution of Julianne Lynch in the preparation of this chapter are gratefully acknowledged.

REFERENCES

Department of Labour. (1989). *Maths multiplies your choices*. Author: Melbourne.

Faludi, S. (1991). *Backlash*. New York: Crown.

Kenway, J., & Modra, H. (1992). Feminist pedagogy and emancipatory possibilities. In C. Luke & S. Gore (Eds.), *Feminisms and critical pedagogy* (pp. 138–166). New York: Routledge.

Kristeva, J. (1981). Women's time. *Signs: Journal of Women in Culture and Society*, *2*(1), 13–35.

Leder, G. C. (1994). Using research productively. In C. Beesey & D. Rasmussen (Eds.), *Mathematics without limits* (pp. 67–72). Brunswick, Vic.: Mathematical Association of Victoria.

Mura, R. (1992). *Feminist theories underlying strategies for redressing gender imbalance in mathematics*. Paper presented at the 7th International Congress on Mathematical Education, Québec, August 17–23.

New South Wales Anti–Discrimination Board. (1984). *Discrimination and religious conviction*. Sydney: Author.

Noddings, N. (1990). Feminist critiques in the professions. In C. B. Cazden (Ed.), *Review of research in education* (Vol. 16, pp. 393–424). Washington, DC: American Educational Research Association.

Offen, K. (1988). Defining feminism: A comparative historical approach. *Signs: Journal of Women in Culture and Society*, *14*(1), 138–166.

Secada, W. G. (1992). Race, ethnicity, social class, language and achievement in mathematics. In D. A. Grouws (Ed.), *Handbook of research on mathematics teaching and learning* (pp. 623–660). New York: MacMillan.

Willis, S., & Kenway, J. (1986). On overcoming sexism in schooling: To marginalize or mainstream. *Australian Journal of Education*, *30*, 132–149.

Windschuttle, K. (1985). *The media*. Melbourne: Penguin.

HELGA JUNGWIRTH

SYMBOLIC INTERACTIONISM AND ETHNOMETHODOLOGY AS A THEORETICAL FRAMEWORK FOR THE RESEARCH ON GENDER AND MATHEMATICS

Reflection on the foundations of research on gender and mathematics shows that there is a clear tendency to trace back gender-specific differences to education and social conditions. The central idea is that girls learn — by the (re)actions of their socializing agents in the context of their mathematics learning and by the segregation of society into female and male domains — that a (successful) occupation in mathematics contradicts femininity, and therefore they turn away from mathematics much more than boys do. Within this very general framework, different views are taken. The preferred approach (cf Menacher, 1994) is the social-psychological one; in empirical studies mainly concepts stemming from social psychology[1] are used, and primarily the following ones:

- attribution theory — this is the basis when gender-specific patterns of the attribution of success and failure are thematized or when female underachievement in mathematics is explained by the fear of success of women in such male domains
- self-concept — this is the theoretical background for the analysis of differences in confidence in learning mathematics
- concept of attitude
- role concept — this is used when the girls' relative avoidance of mathematics is ascribed to the lack of appropriate role-models

In addition, within research on gender and mathematics, specific concepts for the explanation of gender-related differences in the relationship to mathematics have been developed — for example, to mention only two models of particular importance, the Autonomous Learning Behavior (cf Fennema, 1989) or the Model for Academic Choice (Parsons et al., 1983).

The basic idea of the preferred social-psychological approach, whatever concepts are emphasized in detail, is to understand respectively the girls' and the boys' relationship towards mathematics as a product of certain factors which act upon them and have the effects observed. Therefore, many studies are interested in isolating factors to which the analysed aspects of the relationship to mathematics can be ascribed.

A look at the methods and the methodology in the research on gender

G. Hanna (ed.), Towards Gender Equity in Mathematics Education, 49–70.

and mathematics shows a preference of natural science oriented procedures, although this preference has decreased quite recently. The classical procedure starts with rendering the phenomena under question measurable, formulating a hypothesis about the behavior of the quantities to be measured, choosing a random sample, then testing the hypothesis in this sample. Studies designed in this way try to arrive at statements of result which are as universally valid as possible — that is, as the random sample provides. The fact that the design outlined is widespread within research on gender and mathematics is not surprising, as the disciplines it recurs on — social psychology, as already noted, and other branches of psychology and pedagogy — are totally or at least partly oriented to natural science and its methodology. As Ulich (1976, p. 29) says with respect to social psychology: "[The goal is] to investigate invariable accounts that one hopes to discover in the timelessness and unchangeableness of attributes, events, and relations" (my translation). Thus the research on gender and mathematics takes over the perspective of the background disciplines. Fennema (1993, p. 13) sums this up: "Much of what we know about gender and mathematics has been derived from scholarship that has been conducted using a traditional social science perspective — i. e., a positivist approach."

Although research on gender and mathematics has been very successful — after all a lot of important insights in the relation between mathematics and gender have been gained — I would like to plead for an extending of the background of thinking leading this research by conceptualizing the phenomena in certain ways and by identifying the proper ways to analyse them empirically. In my (but not only in my) opinion, within the usual mental framework, important questions cannot be investigated. Fennema (1993) in her report about the state of the art speaks for a new scholarship on gender and mathematics and discusses two promising perspectives for research on gender and mathematics: the cognitive science and feminist perspectives. Cognitive science analyses, among other things, subjective theories about mathematics belief systems or the thinking about specific mathematical domains. It might be that the two genders differ in these aspects, and these differences could be of great importance for the understanding of the gendered relationship to mathematics. From the feminist point of view, gender is a basic category for understanding our approaches to the world — our perceptions and beliefs and scientific findings as well. In our society, male perspectives dominate. More feminist scholarship (cf. for research done under this perspective Burton, 1986; Damarin, 1990; Jacobs, 1994) could be fruitful for research on gender and mathematics because the revelation of bias in favor of the males (cf Kimball, 1993) within mathematics education, questioning even the neutrality of mathematics itself, and the development of female-adequate forms may take on a new quality.

I agree with Elizabeth Fennema that the dominating social science research perspective limits the understanding of gender and mathematics. In my opinion, a further aspect she does not mention has not yet received appropriate attention — that of the everyday creation of what we call the typical female or male relationship to mathematics through a multitude of subtle processes in the family, in the mathematics classroom, among peers, among adults in the workplace, in everyday affairs.

Let me give an example of what I mean. We have rather strong evidence that parents are an important factor for the development of a close relationship with mathematics. But what really makes an encouraging climate in a family?[2] What happens day by day? What do the mothers do, what do the fathers? Do the girls — and the boys too — contribute to parents' believing their children to be qualified for mathematics and to supporting their mathematics-related aspirations? Is the girls talking about mathematics and their learning of mathematics perhaps contradictory in a way, more likely to be misunderstood, so that parents tend to imagine problems with mathematics on the part of their daughters rather than on that of their sons? Within the dominating framework of social psychology with its natural science oriented methods, such questions hardly can be analysed adequately, as the goal is the isolation of influencing factors but not a thick description of the family climate overall. For example (cf Parsons, Adler, & Kaczala, 1982) mothers' perceptions of the importance of mathematics for their daughters in comparison with their sons is isolated as a variable and connected with the daughters' and sons' attitude towards mathematics. The process itself, in the sense of a network of mutually related actions creating the mathematics-related self-concept and that about others, is out of scientific reach; and the question about the girls' and the boys' contribution to the family climate concerning mathematics probably seems absurd when an influence in one direction only — from the environment on the girls and boys — is assumed.

To sum up: in order to make it possible to analyse the development of the two genders' relationships to mathematics context-bound and process-bound, another theoretical perspective and — connected with this — other methods are necessary. Adequate theoretical frameworks for this undertaking are symbolic interactionism and ethnomethodology. In the following, I first want to present the main features of these approaches and then work out the significance for the gender and mathematics issue. Both — it should be mentioned — are well-established perspectives within research on mathematics education not concerned with the gender question.

SYMBOLIC INTERACTIONISM AND ETHNOMETHODOLOGY — A SHORT SURVEY

Symbolic interactionism (cf. Blumer, 1969; Mead, 1934) can be characterized as an approach to interpersonal acting. The expression "symbolic" shows that in this approach action is taken into a consideration that involves interpretation of previous action, or in the terms of Mead, that uses "significant symbols." It is not denied that people also interact in a non-symbolic way — for instance when somebody automatically reacts to a bodily movement of the vis-à-vis — but the mode of interaction considered characteristic for human beings is symbolic interaction, as human society is directed, in principle, to mutual understanding. Symbolic interactionism is based on three presuppositions:

> The first premise is that human beings act toward things on the basis of the meanings that the things have for them. Such things include everything that the human being may note in his world — physical objects, such as trees or chairs; other human beings, such as a mother or a store clerk; categories of human beings, such as friends or enemies; institutions, as a school or a government; guiding ideas, such as individual independence or honesty; activities of others, such as their commands or requests; and such situations as an individual encounters in his daily life. The second premise is that the meaning of such things is derived from, or arises out of, the social interaction one has with one's fellows. The third premise is that these meanings are handled in, and modified through, an interpretative process used by the person in dealing with the things he encounters. (Blumer, 1969, p. 2)

The first presupposition itself is not specific to symbolic interactionism. All constructivist concepts share this assumption, as for example cognitive science when saying that behavior is guided by mental activities or cognitions. More essential are the second and the third presuppositions — they only legitimate the term "interactionism" as through them interaction gets its specific importance. The second premise separates symbolic interactionism from cognitive-oriented approaches which consider the meaning a thing has for a person to be a result of the given psychological elements of this person — like her or his feelings or memories. Finally, the third presupposition differentiates between symbolic interactionism and concepts working on the principle that the meaning of a thing is formed once and for all and that the use of meaning is only an application of this meaning. This third premise does not exclude a stabilization or reinforcement of meaning, but these are the result of a corresponding process.

Starting from these premises, symbolic interactionism works out concepts about human group life and human behavior. For the purpose of this chapter, it is important to say that "action on part of a human being consists of taking account of various things that he notes and forging a line of conduct on the basis of how he interprets them" (Blumer, 1969, p. 15). Human group life is considered a process in which people indicate to others their lines of conduct, interpret the signs made by others, and adapt their lines of action to one another. People inter-act in the literal sense of the word. Summing up one could say that interaction is an interpretative process. People show each other the meanings of the things, adjust these meanings according to one another's reactions, stabilizing or modifying them at the same time. Commonly accepted meanings — in traditional terms: valid meanings — are the result of social negotiation processes. Whatever course an interaction takes, and whatever the results are of an interaction, it is all the product of the acting of its participants. This statement includes two aspects.

First, what happens in an interaction is not just the "sum" of the behavior of the participants concerning the factors considered to direct the behavior. The course of interaction is determined neither by interpersonal traits of the people involved — like attitudes, cognitions, or needs — nor by external facts — like social norms or role stereotypes. It is not considered to be analogous to a physical process, calculable and entirely predictable from the initial state. "It is of great importance that a participant for him or herself cannot completely determine the form or the effect of his or her utterances, and even less the course the interaction takes but, on the contrary, that in the course of the interaction a structure emerges which none of the participants would have been able to create alone" (Auwärter & Kirsch, 1982, p. 273; my translation).

Second, all participants in an interaction contribute to what happens in their interaction. They all influence the way it develops. This does not mean that the result would have to mirror each line of conduct or each perspective equally. Saying that all take part does not indicate a balance of all intentions, interests, and so on. The position that all participants of an interaction contribute their share to its course and its result applies as well when — from the participants' point of view, or from that of an observer — only one participant has her or his way. That she or he can do this does not only depend on her or him; succeeding in one's intentions is made possible by the others too.

The second approach I want to outline briefly is ethnomethodology (Garfinkel, 1967; Leiter, 1980). The term was minted by Garfinkel and it is a branch of phenomenologically oriented sociology. Like symbolic interactionism, ethnomethodology assumes that the world we live in is established by the mutually related acting of the members of society. What people in their everyday activities perceive and treat as objective facts, as a reality existing independently of their interventions, they establish by their perceptions and actions. Social reality is not something lying behind or beyond the everyday behavior nor something determining this behavior. Social reality is made into reality in the course of action.

While symbolic interactionism develops this core idea by giving descriptions of human group life and human behavior, based on the concept of interaction presented above, ethnomethodology elucidates the fact that this aspect of establishing social reality generally is not seen; social reality, on the contrary, is taken as given, external reality. Ethnomethodologists say that this comes about because, parallel to the managing of the everyday affairs, they are described and explained. Social reality is established by activities indicating the meaning of what is done: "The activities whereby members produce and manage settings of organized everyday affairs are identical with members' procedures for making those settings 'accountable.' ... By his accounting practices the member makes familiar, commonplace activities of everyday life as familiar, commonplace activities"

(Garfinkel, 1967, pp. 1, 9). The goal of ethnomethodological analyses is to work out how this is done — that is, to identify the methods that the members of society use to establish the meaning the world has for them. The making of social reality — ethnomethodology says — cannot be done in quite subjective, irregular ways because members of society or subgroups of them take part in these processes. They must proceed somehow methodically, regularly, and these methods are in the centre of interest of ethnomethodology. For example Sacks (1974) analyses the way a joke is made a joke by the joke-teller and her or his audience. Garfinkel (1967) inquires into the methods a male-to-female transsexual uses to make herself be seen as a woman.

Thus both approaches conceptualize the world as one established by the mutually related actions of human beings. Symbolic interactionism directs attention to people's being related to one another by interactions, whereby interaction is seen as an interpretative process which the meaning of things arises out of and is handled in. Ethnomethodology concentrates on the methods by which the external social world is established as this world.

THE INTERACTIONIST AND ETHNOMETHODOLOGICAL PERSPECTIVE ON MATHEMATICS EDUCATION

Symbolic interactionism and ethnomethodology offer a perspective on the world that can be made fruitful in the pedagogical and didactic context. Because of its generality, this perspective does not lead to a specific research program; its application, however, means to emphasize a certain aspect of the subject of pedagogy and didactics — that education, respectively teaching and learning, is an interactive process and that the results are constituted interactively. Pedagogy and didactics guided by symbolic interactionism and ethnomethodology take place within the "transformation-oriented paradigm" (Ulich, 1976). In this paradigm, regularities are ascribed to conditions which are changeable in principle because they are considered to have been established.

Within mathematics didactics, symbolic interactionism and ethnomethodology in the meantime are well-accepted background theories for the analysis of the teaching and learning processes. These approaches are a sound basis for inquiring into the social dimension of the teaching and learning of mathematics. To neglect this dimension is seen as a shortcoming, and there is a growing number of studies which take this dimension into account.

From the perspective of symbolic interactionism and ethnomethodology, the teaching and learning of mathematics are seen as an interpretative process as well. The participants — the teacher and the students — construct subjective meanings from the themes and events in the mathematics classroom. This means among other things that there is no immutable objective mathematics. The shared mathematical meaning of the objects dealt with is negotiated in the interaction. Both the teacher and the students take part in the negotiation process. Both bring in their subjective views.

In general, the teacher and the students differ in their views. The teacher looks at the mathematical issue from the perspective of the discipline of mathematics, and teaching itself is probably oriented by pedagogical concepts about an effective mathematics education or by the perceived social function of school. For the students, on the other hand, the everyday experiences are significant when looking at the mathematical content of a lesson and at what was done and how it was done in the lesson before. Thus the teacher and the students — and during group-work the students among one another — have to achieve understanding. They must come to an agreement as to what the official meaning should be. From this perspective it is obvious that there may be misunderstanding between teacher and students as well as between the students. Students could make "mistakes," and the teacher's efforts may be in vain.

Starting from this basis, researchers in mathematics education investigate very different problems. Cobb, Wood, and Yackel (1991, 1992) analyse how the change of social norms in the mathematics classroom with respect to the participation in the classroom discourse (stressing group-work, mathematical reasoning, acceptance of failure) affects the students' learning. Bauersfeld, Krummheuer, and Voigt (1988) investigate the negotiation of mathematical meaning and the experiences teacher and students have in this process. Bauersfeld (1980, 1988, in press) has developed the concept of "subjective domain experiences" and analyses among other things interaction relating to the notion of language game. Krummheuer (1982, 1992, 1993) has analysed — using the concept of framework by Goffman (1974) — background understandings on the part of the teachers and the students and deals with the question of how collective argumentation in elementary schools is organized. Voigt (1984, 1989, 1994) has reconstructed hidden routines on part of teachers and students and patterns of interaction in which the fragility and the trouble proneness of the negotiation of mathematical meaning is reduced. He is interested in the ways the school-mathematical view of the topics of the elementary mathematical classroom is constituted. Steinbring (1991) has reconstructed the epistemological structure of mathematical knowledge that is established in the teacher-student interaction. Apart from the researchers named above, there are several scholars whose studies show that they share the perspective on mathematics education induced by symbolic interactionism and ethnomethodology, although they do not refer to these approaches explicitly (Forman, 1989; Maier, 1991; Saxe, 1990).

METHODOLOGICAL IMPLICATIONS OF SYMBOLIC
INTERACTIONISM AND ETHNOMETHODOLOGY

Before trying to demonstrate the specific value of symbolic interactionism and ethnomethodology for research on gender and mathematics, methodological consequences shall be touched upon.

The assumption that people establish the meaning the world holds for them within their common, mutually related acts implies that gaining insights in this world and in its making needs an "open" approach on the part of the researcher that permits perceiving the world from the perspective of the people involved. This does not mean that findings could comprise only what the people themselves know and say. The researcher, too, interprets what she or he observes; she or he constructs meaning, too — of the meaning indicated to her or him — and this she or he does with the particular view leading her or his research. In principle, researchers are in the same position as the people whose world they want to investigate; people indicate to others, and to the researchers as well, their expectations, lines of conduct, orientations; their utterances are often ambiguous and can only be understood in context. People, and the researchers as well, decipher them by the method of "documentary interpretation." To apply this method

> means that a pattern is identified which is at the root of a number of phenomena. Thereby each phenomenon is considered as related to the underlying pattern — as an expression, as a "document" of this pattern. This, on the other hand, is identified by the various concrete phenomena so that the phenomena and the pattern itself determine each other. (Wilson, 1981, p. 60; my translation).

Unlike the everyday interpreting, within research there are certain standards scientific interpreting has to meet (cf Beck & Maier, 1994; Voigt 1990).

It is obvious that the method to be selected for a particular investigation is not yet determined by these general methodological considerations; it depends on the question to be analysed. In any case, however, it is important to take into account that the meaning of things depends on the context and is tied to the process it emerges from. Comparing empirical research based on symbolic interactionism and ethnomethodology with other investigations one could say that it is not designed like a controlled experiment in natural science with its mathematical-statistical analysis — which is often the model for empirical studies in psychology and social sciences — but rather like an exploratory trip to an unknown culture. It was not without good reason that within mathematics didactics empirical work based on symbolic interactionism and ethnomethodology has been labelled "microethnographic."

WHAT COULD BE THE PRACTICAL USE OF SYMBOLIC INTERACTIONISM AND ETHNOMETHODOLOGY FOR RESEARCH ON GENDER AND MATHEMATICS?

I want to focus on three aspects which appear to me particularly important. They are not specific to symbolic interactionism and ethnomethodology in the sense that these and only these approaches would emphasize them (see the references below), but they are essential for them so that through using them as a basis for research on gender and mathematics, these aspects would gain in significance. Claiming them from other points of view is supported as well.

Emphasis on an Active Role for Females

From the interactionist and ethnomethodological perspective, females play an active role in the development of the typical female relationship to mathematics. They do their share in making this relationship having the features observed. It is important what females do or do not do: how they deal with mathematics, how they talk about mathematics — in the family, among peers, in the classroom, in everyday situations. In all this they make indications to their fellows, and these are taken up and interpreted. Whether the meaning read into the females' math-related actions and speech, however, is the same as the meaning intended by them is an open question. There are even theoretical perspectives which, although on different levels, presuppose a systematic misunderstanding of women respectively between females and males (Gilligan, 1982; Maltz & Borker, 1982; Tannen, 1990). Yet the assertion is not that females are the only ones responsible for the status quo of the female relationship to mathematics but that they contribute to this relationship's being as it is.

Within research on gender and mathematics, this aspect is often suppressed. If the common social-psychological view — which sees only an influence from the environment on the girls — is taken into account, it is not surprising that they are conceptualized as passive, reactive beings. In principle, from the social-psychological point of view, males as well are molded by the environment; in spite of this, they are considered at least partly going their own way. The girls, or women, on the contrary, are the victims of the status quo, not contributors, much less authors.

An example showing very clearly the ignorance of females' contributions to the situation are studies dealing with the interaction in the mathematics classroom. Such studies are usually interested in working out the treatment of the girls (and boys) by the teachers. This can be seen even in the design of these studies; there are many more categories provided to cover the teacher's behavior than that of the (female) students (cf Becker, 1981; Eccles & Blumenfeld, 1985; Fennema & Peterson, 1986; Leder, 1990; Stallings 1985). This can also be seen from the discussion of the results.

Special emphasis is placed on the different treatment teachers give to girls and boys — they interact in many categories more frequently with boys than with girls — but the treatment is hardly related to the behavior of the girls themselves. Such studies concede, at the most, boys having an influence on the behavior of their teachers: They get more disciplinary attention because they disturb teaching more than girls do. But when the girls take part in the classroom discourse less frequently than boys — as several studies show (Leder, 1990; Stallings, 1985; for a survey see Meyer & Fennema, 1988, too) — it would make sense to connect this result with the teachers' behavior towards the girls and to classify the latter as a reaction to the reservedness of the girls. This is not done or if it is, not in the way that the girls are really considered active, their teachers' behavior influencing participants in classroom interaction. Perhaps Eccles and Blumenfeld (1985) mean this but they say only that quite generally differences in girl-teacher interactions and boy-teacher interactions in their study are due to gender-specific differences on the part of the students just as much as to prejudices on the part of the teachers. Fennema (1993, p. 20) says — in another context, however — that "teachers appear to react to pressure from students, and they get more pressure from boys," which is an indirect statement about the girls' contribution to the status quo. German studies done by feminist educators (Dick, 1992; Enders-Dragässer & Fuchs, 1989) in principle admit that girls contribute to what happens in the classroom; they as well, however, perceive only girls' facilitating "normal" teaching by their disciplined participation. The view that they could influence the academic discourse too, and their teachers' actions in this discourse, cannot be reconstructed from their writing.

I want to demonstrate now by an example taken from a mathematics classroom that from the interactionist and ethnomethodological perspective, girls absolutely influence the course of the academic interaction. The scene presented below took place in a Grade 8 class (students aged about 14) at an Austrian grammar school. The class is learning about the Pythagorean theorem and its applications. The results of the homework are discussed. One task went as follows: "Given are two line segments with length a and b. Construct a segment of a line with the length $\sqrt{a^2 + 4b^2}$." Julie said that she had plotted "a and four times b" and was called to the blackboard in order to calculate the hypotenuse from her cathetes. The teacher intended her to realize that her solution could not be right and wanted her to correct it. The hypotenuse, however, she calculated was $\sqrt{a^2 + 4b^2}$. Then the following dialogue began (for the rules of transcription see Appendix):

18 Teacher: well a moment yes´ tell me the Pythagorean theorem in
19   words what does it describe how can we interpret it
20   geometrically- (draws a square in the air with his hands)
21 Julie: well
22   the two cathetes *which* do the right do the right

23 Teacher:	nono. nono.
24	what did we do in the proof. (2 sec p) (draws squares with
25	his right forefinger at the cathetes of the triangle sketched
26	on the blackboard)
27 Julie:	*well* we always extended the *expressions´*
28 Teacher:	no´
29 Julie:	(*by?*)
30 Teacher:	no I am driving at something else- (Sam raises his hand)

In this scene the answers of the girl are rejected by the teacher more and more explicitly. While at the beginning, he just puts the solution aside without calling it wrong and then rejects her suggestions, at the end of this dialogue, he definitely states that he is aiming at another answer. In addition to the fact that her homework is wrong now it becomes evident that Julie does not know the geometrical interpretation of the Pythagorean theorem. What happens can be called an "emerging of failure" (Jungwirth, 1991a). This, one might argue, is due to the teacher — to his approach to her wrong solution and to his questions. But is he really to blame for this emerging of failure? From the perspective of symbolic interactionism and ethnomethodology, he is not. She contributes a good measure to what happens as she persistently ignores his hints and goes her own way. Because she does not take up his hints — drawing her attention to the proof can be seen as an additional attempt to guide her — there is practically nothing else the teacher can do but to reject her answer quite explicitly, at least not in the situation where wrong homework should be corrected in the usual way, which means as quickly as possible but nevertheless making the student understand what was wrong (cf Jungwirth, 1993; Voigt 1984, 1989).

The basic idea I want to present is that from the perspective of symbolic interactionism and ethnomethodology, females are active beings not only influencing the course of interaction in the mathematics classroom but quite generally contributing their part to our world being as it is.

It should be noted that not only these approaches take this view. Within the German feminist movement, for example, Thürmer-Rohr strongly criticizes the victim concept:

> Feminists again and again explain the invisibility of women in history and their relative ineffectiveness in the present time by saying: Women *were* not permitted to develop independence, freedom, confidence in themselves, and competence ... women *are* obstructed.... To see female behavior and female existence as always passive is a discrimination against ourselves. (Thürmer-Rohr 1992, p. 50ff; my translation).

As an alternative, she develops the concept of "complicity of women." It says that our patriarchal society would not exist as it does unless women had contributed their part to its existence.

Taking into account that asking the gender question always — in the research on gender and mathematics, too — includes being critical of the marginalization of women, the consideration of the contribution of women to the status quo is nothing but consequent. Not thematizing this contribution leads to a paradoxical situation: On the one hand, overlooking the experiences, the achievements, the actions, the *weltanschaungen* of females is an offence, and on the other hand, however, this overlooking is reproduced within research by considering the females' contribution to what is and how it is — for example to the girl-teacher interaction — as of no importance.

Emphasis on the Interactive Creation of So-Called Facts

Symbolic interactionism and ethnomethodology direct attention to the creation of the relationship of women (and men) to mathematics. The focus is on the processes, the ways in which the phenomena we know are established.

It may be that this is not seen as a new quality that symbolic interactionism and ethnomethodology can bring in, as it is in general the goal of research on gender to reveal how gender-related differences in the relationships to mathematics come into being. Usually such a revelation means explaining the differences by factors considered effective because of statistical calculations. The factors themselves are understood as given or they are traced back to underlying factors until the basis seems unquestionable. The most general bases are gender-role stereotypes. In the end, it is the feminine stereotype that makes women turn away from mathematics: It results in "wrong" expectations on the part of parents, teachers, and so on, thus in "wrong" mathematics-related treatment of girls, in "wrong" role models for girls, and makes the girls themselves believe that a (successful) study of mathematics contradicts femininity.

The interactionist and ethnomethodological view as well do not deny that there are gender-role stereotypes in our society and that they are models for people's actions and for their interpreting the actions of others. What makes the difference, however, is that even the stereotypes are not seen as concepts which, once established, survive and always will be effective but as concepts which can survive only if they are reproduced in actions day by day. They are not only the cause of action but also its consequence.

The ethnomethodologists Kessler and McKenna (1978) transcend this position a bit by saying that gender itself — that is, our concept of two genders — is a social construct. They believe:

> Gender attribution is a complex, interactive process involving the person making the attribution and the person she/he is making the attribution about.... The process results in the "obvious" fact of the other being either male or female. On the one hand, the other person presents her or himself in such a way as to convey the proper cues to the person making the attribution.... The second factor in the interaction are the rules (methods) that the person doing the attributing uses for assessing these cues. These rules are not as simple as learned probabilities, such as people with beards are usually men. They are rules that construct for us a world of two genders.... Part of

being a socialized member of a group is knowing the rules for giving acceptable evidence for categorizing. (Kessler & McKenna, 1978, p. 6).

Four areas of self-presentation, especially the first two, are important for gender attribution: "(1) general talk (both what is said and *how* it is done), (2) public physical appearance, (3) the private body, (4) talk about the personal past" (Kessler & McKenna, 1978, p. 127). That biology plays such a prominent role is not due to nature but to culture: "Our reality is constructed in such a way that biology is seen as the ultimate truth" (Kessler & McKenna, 1978, p. 162). The social construction of the two genders is the basis for all scientific work dealing with gender-related differences.

I want to give an example now, taken from a study on gender and mathematics, of perceiving a process instead of seeing a product.

AN EXAMPLE: THE INTERACTIVE ESTABLISHING OF MATHEMATICAL COMPETENCE

It is quite usual to understand the mathematical competence of a student as a property of this student, becoming evident in her or his giving correct answers, bringing clever arguments, asking intelligent question, making well-considered suggestions, and so on. By all this the student indicates that she or he has a good mathematical knowledge or understanding — or in short: that she or he is competent in mathematics. This may have several reasons: cognitive prerequisites, affective factors such as the confidence in learning mathematics, and so on. From the perspective of symbolic interactionism and ethnomethodology, however, the mathematical competence of a student expresses the communicative situation as it is at the actual time (cf Jungwirth, 1991b). Mathematical competence is the result of a social interaction. All participants — not only the student but the teacher as well — contribute to the development of such a competence. It comes into being by the mutually related action. This view modifies the concept of mathematical competence. Good mathematical knowledge and understanding are no longer decisive; a critical element as well is the ability to present this knowledge and understanding adequately. Mathematical competence of a student is established by certain practices used by the student and the teacher in the classroom. Of course it is an open question whether competence in this sense always means competence in the classical sense. How does mathematical competence of students develop in detail? It arises out of an interaction which — in the "natural attitude" (Schütz, 1971) — is smooth running. Such a smooth running is the result of an action of a certain type being followed by an action of the corresponding type. Linguistically speaking, the "implicit obligations" (Kallmeyer & Schütze, 1976; Voigt 1989) are satisfied; seen psychologically, smooth running develops when all participants in an interaction act in such a way that they satisfy the expectations directed at them by their fellows through their own actions. Or in other words, when the student's actions correspond to the teacher's, and vice versa.

By the following two examples, this view of mathematical competence is illustrated. In the first example, the mathematical competence of the student is established in the teacher-student interaction, in the second, this does not work as it could.

The first example is a dialogue that took place in a Grade 10 class (students aged about 16). The parabola $y = x^2$ and its graph have been introduced, and then the parabola $y = a(x-c)^2 + d$ and its graph. The class repeats how the graph is modified by a, c, d. The effects of factor a have been discussed already (for the rules of transcription see Appendix):

01 Teacher: well which other modifications are still missing. Arthur.

02 Arthur: *well* if a d, is put to it for instance´

03 Teacher: well to say if a number is added what happens then´

04 Arthur: then the the vertex I mean the the the the origin

05 Teacher: the parabola yes´

09 Arthur: the parabola is simply shifted on the hori on the vertical

10   axis´

11 Teacher: is shifted upwards or downwards yes´

12 Arthur: *and* if the d is **greater** than zero then it is shifted upwards´ if

13 Teacher: yes´

14 Arthur: it is less than zero it is shifted downwards.

15 Teacher: right. and , there is a final modification-

By which methods is mathematical competence of Arthur established? Focussing first on his utterances, the relative taciturnity is striking. He uses the routine of "verbal reduction" (Voigt, 1989), that is he does not say more than absolutely necessary in the actual moment. For example, in his first utterance, Arthur only considers that d is added but does not name the mathematical operation (addition) nor does he say anything about the effect. With this practice he tries to minimize the risk of showing a lack of knowledge or understanding. The (female) teacher, on the other hand, contributes to the establishing of the mathematical competence as she satisfies the implicit obligations of his practice. By this the vis-à-vis is invited to elicit the aspects still missing or to complete the fragments to a correct answer. She does both. By her questions and her signals confirming the fragmentary answers and requesting continuation (see, for example, the "yes" in line 5), she elicits the missing aspects. As she decomposes the development of the solution into small subsequent steps (cf Voigt, 1989, for this routine) the student, on the other hand, is not obliged to say more than he actually does in his utterances. Second, she turns his indications into correct answers by another routine called "teacher's echo." Again the obligations can be seen the other way round: as it is she who fixes the meaning, the student is only obliged to make indications. The teacher's acting and the student's acting correspond with each other; there is a smooth interaction that the mathematical competence of Arthur emerges from.

The second example is taken from a Grade 8 class (the same which the scene with Julie derived from) at an Austrian grammar school. The class learns about the Pythagorean theorem and its applications. The results of the homework assignment are discussed. One task was to trace the square root of 5 using the Pythagorean theorem. Before the following dialogue started, the teacher called upon a student who outlined the technical procedure ("first draw one cathetus then the other one in a right angle to the first") but did not know the necessary decomposition of the square root of 5 and gave a wrong value for one cathetus — five instead of two.

23 Teacher:	which square root should you draw. what is it.
24 Bill:	the square root of five
25 Teacher:	the square root of five. well how can we do this in a **quite**
26	simple way- (Albert, Harry, Max, Peter, Sam, Sarah, Vicky
27	raise their hands) (4 sec p) Mary have you got it´
28 Mary:	yes´ (first I?)
29 Teacher:	how did you do it´ (Albert, Harry, Max, Sarah, Vicky put
30	down their hands)
31 Mary:	well the square root of two square plus one square. (Peter
32	and Sam put down their hands) that is , well you first have
33	marked off two and one. and thus the the , the line segment
34	is (just?) two.
35 Teacher:	well once again. the one cathetus is **how** long-
36 Mary:	two and
37 Teacher:	the other cathetus is-
38 Mary:	and is one.
39 Teacher:	then the hypotenuse we get is the square root of five. isn't
40	it´ a square plus b square.

Why is mathematical competence on the part of Mary not really established in this dialogue? Mary gives a complete answer — for example, an answer in which she describes the main aspects left to say. By this practice, she departs from the usual way of answering. Her action does not include the obligation for the teacher to elicit the missing aspects by further questions. In spite of this, the teacher acts routinely in this manner. He turns to the different aspects of her answer and asks about them one after the other. The answers on these restricted questions in sum correspond with the answer already given. The teacher's reaction can be called an "undoing" of the complete answer. It is an attempt to re-establish the common step-by-step development of solutions the girl has disturbed.[3] The type of action of the girl and that of the teacher do not correspond with each other. The effect is that she is relatively incompetent: Her answer might have been correct in its core but — seen from the reaction of the teacher — she spoke gibberish and was not able to make herself understood.

What is not discussed further in this context is the motive of the teacher for undoing the complete answer. It maybe that in his opinion Mary's answer was just too dense for the other students; however, he does not mention this, and from the perspective adopted here, only those aspects that are of importance are indicated in the vis-à-vis.

It is not without good reasons that in the example illustrating the establishing of mathematical competence the student is a boy, and in the other one a girl. As a result of my investigation of interaction in the mathematics classroom (Jungwirth, 1991a, 1991b, 1993), successful — in the sense of corresponding with the teacher's actions — participation in the classroom discourse as it usually goes has become a routine more for boys than for girls. Boy-teacher interactions tend to run smoother than girl-teacher interactions — not always, but if there is a mismatch in the academic interaction then it is with girls. Consequently, with boys more mathematical competence is established.

The interactionist and ethnomethodological approach to mathematical competence offers a new explanation for the difference in teachers' perceptions of the learning of mathematics of girls and boys as well. Studies say that teachers tend to use such terms as "natural abilities" or "flair" much more often with boys whereas they call even high-ranking girls "hard-working" (cf Fennema, Peterson, Carpenter, & Lubinski, 1990; Walkerdine, 1989). This is often understood as a prejudice against girls because of internalized gender-role stereotypes. The statements of the teachers, however, could reflect just the differences in the interactive mathematical competence. If girls do not participate in the classroom discourse as successfully as boys but get better marks in the exams then it seems very likely for the teachers to assume that girls are more diligent and hard-working.

To understand that what seems to be a fact is being established day by day may be fruitful not only for the analysis of the status quo but for its transformation as well, since, from this perspective, anonymous social powers obstructing changes can become conditions created on the spot, and the establishing of these may be influenced. For example: those who on principle attribute an underestimation of the girls' mathematical achievement to the general conditioning of the teachers according to the female-role stereotype find themselves — when trying to alter the estimation of the girls — confronted with a very powerful opponent — with deep-rooted, omnipresent gender-role stereotypes. To struggle against these is hard, and besides the everyday experience seems to confirm the belief again and again. If, however, the establishing of mathematical competence on the spot is seen, or its failing with girls, the task is "only" to deal with the concrete situation and to try to find a new way of repeating a lesson or of correcting wrong solutions. I do not want to deny that this could be hard work too because teacher and students have to change interaction practices they are used to, but the problem can be localized much better; the "opponent" is much more tangible.

Emphasis on Heterogeneity Within the Genders and Their Learning of Mathematics

Symbolic interactionism and ethnomethodology emphasize the possibility of heterogeneity within the genders' relationships to mathematics and their learning of mathematics. Interactively negotiating the meaning of the things, and by this establishing social reality, implies that this may be done in different ways and that it may lead to different results. This means that a homogenous reality cannot be expected a priori. Uniformity may develop but it is not necessarily the case. Differences — whatever the subject under consideration is — are always possible.

It may be that this position does not appear as a specific feature of symbolic interactionism and ethnomethodology. Indeed, the pleading for the possibility of diversity within the two genders, in particular within the female gender, is not characteristic of these approaches only. Feminists, especially deconstructive feminists (cf de Lauretis 1986, Moi 1989, Schor 1989), criticize quite generally concepts of women — regardless of whether they see the female founded in a natural essence or in a character that developed socially — that define them just as the opposite of men without taking into account the manifold differences among women.

With respect to the issue of this article, this position means that within each gender, mathematics-related differences may occur, respectively within the teaching and learning of mathematics of each gender, or more precisely, that such differences are theoretically permissible. The opinion dominating within the scientific community, namely that the development of the females' and males' relationship to mathematics is due to quite a number of factors, would suggest that statements about this issue are non-uniform, even contradictory. There is some evidence that permitting heterogeneity is "realistic" in a way. An example: until now a clear-cut relation of gender-related differences in teacher-student interaction and gender-related differences in attitudes and achievements has not been demonstrated (cf Eccles & Blumenfeld, 1985; Koehler, 1990; Leder, 1990). Of course by pointing out the possibility of heterogeneity, I do not mean that results being obtained again and again would be suspicious.

The significance of this third consideration may become even clearer if implications for empirical work are outlined. Seeing the subject of analysis as potentially heterogeneous, the importance of an empirical study cannot be measured by the range of validity of its results. Within classical social-psychological research, however, this is an important criterion. The sector of reality for which its results are valid should be large in order that a study can gain importance. Demanding this indicates seeing the subject of research as homogeneous, as uniformly structured, because it is under this presupposition that it makes sense to aim at statements that are universally valid (cf Ulich, 1976). Theoretically permitting heterogeneity to the subject of research is of practical use for research on gender and mathematics as studies not keeping to the standards derived from natural science can be conducted without having a "guilty conscience," so to speak. And to accept procedures not oriented to the methodology of natural science is important, because this orientation restricts analyses to those phenomena that can be conceptualized in natural science's way, and within the subject of gender and mathematics these are not all of them.

NOTES

1. Provided that there is an appropriate "definition" of social psychology. There are, however, different answers to the question of which concepts do number among those of social psychology (cf Frey & Greif, 1987).

2. There is some evidence that the situation is so complex that it cannot be really represented by isolated measurable quantities. For example, in an Austrian-wide study among female students of mathematics, science, and technology (Jungwirth, 1992) only about 16 percent said that they were encouraged to play with construction sets.

3. This is the usual way, at least in German-speaking countries.

REFERENCES

Auwärter, M., & Kirsch, E. (1982). Zur Entwicklung interaktiver Fähigkeiten — Begegnungskonstitution und Verhaltenssynchronie in der frühen Kindheit. *Zeitschrift für Pädagogik, 28*, 273–298.

Bauersfeld, H. (1980). Hidden dimensions in the so-called reality of a mathematics classroom. *Educational Studies in Mathematics*, 11, 23–41.

Bauersfeld, H.(1988). Interaction, construction, and knowledge — Alternative perspectives for mathematics education. In T. Cooney & D. Grouws D. (Eds.), *Effective mathematics teaching* (pp. 27–46). Reston, VA: National Council of Teachers of Mathematics.

Bauersfeld, H. (in press). "Language games" in mathematics classroom — Their function and the education of teachers. In *Final report of the project the coordination of psychological and sociological analyses in mathematics education*.

Bauersfeld, H., Krummheuer, G., & Voigt, J. (1988). Interactional theory of learning and teaching mathematics and related microethnographical studies. In H. G. Steiner, & A. Vermandel (Eds.), *Foundations and methodology of the discipline mathematics education* (pp. 174–188). Antwerpen: University of Antwerpen.

Beck, C., & Maier, H. (1994). Mathematikdidaktik als Textwissenschaft. Zum Status von Texten als Grundlage empirischer mathematikdidaktischer Forschung. *Journal für Mathematikdidaktik, 15*(1, 2), 35–79.

Becker, J. (1981). Differential treatment of females and males in mathematics classes. *Journal for Research in Mathematics Education, 12*(1), 40–53

Blumer, H. (1969). *Symbolic interactionism. Perspective and method.* Englewood Cliffs, NJ: Prentice-Hall.

Burton, L. (1986). Femmes et mathematiques: y-a-il une intersection? In L. Lafortune (Ed.), *Femmes et mathematiques* (pp. 19–56). Montreal: Les Editions du remue-menage.

Cobb, P., Wood, T., & Yackel, E. (1991). A constructivist approach to second grade mathematics. In E. von Glasersfeld (Ed.), *Constructivism in mathematics education* (pp. 157–176). Dordrecht: Kluwer.

Cobb, P., Wood, T., & Yackel, E. (1992). Interaction and learning in mathematics classroom situations. *Educational Studies in Mathematics, 23*, 99–122.

Damarin, S. (1990). Teaching mathematics: A geminist perspective. In T. J. Cooney & C. R. Hirsh (Eds.), *Teaching and learning mathematics in 1990: 1990 yearbook* (pp. 144–151). Reston, VA: National Council of Teachers of Mathematics.

de Laurentis, T. (Ed.). (1986). *Feminist studies/Critical studies.* Bloomington: Indiana University Press.

Dick, A. (1992). Die Problematik der Koedukation im mathematisch-naturwissenschaftlichen Unterricht. In A. Grabosch, A. Zwölfer (Eds.), *Frauen und Mathematik. Die allmähliche Rückeroberung der Normalität?* (pp. 114–134). Tübingen: Attempto.

Eccles, J. S., Blumenfeld, P. (1985). Classroom experiences and student gender: Are there differences and do they matter? In L. C. Wilkinson & C. B. Marrett (Eds.), *Gender influences in classroom interaction* (pp. 79–115). Orlando: Academic Press.

Enders-Dragässer, U., & Fuchs, C.(1989). *Interaktionen der Geschlechter. Sexismusstrukturen in der Schule.* Weinheim und München: Juventa.

Fennema, E. (1989). The study of affect and mathematics: A proposed generic model for research. In D. B. McLeod & V. M. Adams (Eds.), *Affect and mathematical problem solving: A new perspective* (pp. 205–219). New York: Springer.

Fennema, E. (1993). *Mathematics, gender, and research.* Paper presented at the ICMI Conference, Gender and Mathematics Education, Höör, Sweden, October, 1993.

Fennema E., & Peterson, P. L. (1986). Teacher-student interactions and sex-related differences in learning mathematics. Teaching and Teacher Education, *2*(1), 19–42.

Fennema, E., Peterson, P., Carpenter, T., & Lubinski, C. (1990). Teachers' attributions and beliefs about girls, boys, and mathematics. *Educational Studies in Mathematics*, 21, 55–69.

Forman, E. (1989). The role of peer interaction in the social construction of mathematicsl knowledge. *International Journal of Educational Research, 1*, 55–70.

Frey, D., & Greif, S. (Eds.). (1987). *Sozialpsychologie. Ein Handbuch in Schlüsselbegriffen.* München - Weinheim: Psychologie Verlags Union.

Garfinkel, H. (1967). Studies in ethnomethodology. Englewood Cliffs, NJ: Prentice-Hall.

Gilligan, C. (1982). *In a different voice.* Cambridge, MA: Harvard University Press.

Goffman, E. (1974). *Frame analysis. An essay on the organization of experience.* Cambridge: Harvard University Press.

Jacobs, J. (1994). Feminist pedagogy and mathematics. *Zentralblatt für Didaktik der Mathematik, 26*(1), 12–17.

Jungwirth, H. (1991a). Interaction and gender — Findings of a microethnographical approach to classroom discourse. *Educational Studies in Mathematics, 22,* 263–284.

Jungwirth, H. (1991b). Mathematische Kompetenz als soziales Phänomen. *mathematica didactica, 14*(1), 30–45.

Jungwirth, H. (1992). *Wahl einer mathematisch-naturwissenschaftlich-technischen Studienrichtung und schulische Herkunft bei Frauen.* Linz: Eigenverlag.

Jungwirth, H. (1993). Routines in classroom discourse: An ethnomethodological approach. European Journal of Psychology of Education, *VIII*(4), 375–387.

Kallmeyer, W., & Schütze, F. (1976). Konversationsanalyse. In *Studium Linguistik* (pp. 1–29). Kronberg: Taunus.

Kessler, S., & McKenna, W. (1978). *Gender: An ethnomethodological approach.* Chicago: The University of Chicago Press.

Kimball, M. (1993). *Research perspectives.* Paper presented at the ICMI conference, Gender and Mathematics Education, Höör, Sweden, October, 1993.

Koehler, M. S. (1990). Classrooms, teachers and gender differences in mathematics. In E. Fennema & G. Leder (Eds.), *Mathematics and gender.* New York: Teachers College Press.

Krummheuer, G. (1982). Rahmenanalyse zum Unterricht einer achten Klasse über "Termumformungen." In H. Bauersfeld et al. (Eds.), *IDM- Reihe Untersuchungen zum Mathematikunterricht. Band 5 Analysen zum Unterrichtshandeln* (pp. 41–103). Köln: Aulis.

Krummheuer, G. (1992). Lernen mit "Format." Elemente einer interaktionistischen Lerntheorie. *Diskutiert an Beispielen mathematischen Unterrichts.* Weinheim: Deutscher Studienverlag

Krummheuer, G. (1993). *The ethnography of argumentation* (Occasional paper 148). Bielefeld: IDM.

Leder, G. (1990). Gender and classroom practice. In L. Burton (Ed.), *Gender and mathematics. An international perspective* (pp. 9–20). Strand: Cassell.

Leiter, K. (1980). *A primer on ethnomethodogy.* New York: Oxford University Press.

Maier, H.(1991). Analyse von Schülerverstehen im Unterrichtsgeschehen – Fragestellungen, Verfahren und Beispiele. In H. Maier & J. Voigt (Eds.), *Interpretative Unterrichtsforschung. IDM-Reihe Untersuchungen zum Mathematikunterricht Band 17* (pp. 117-151). Köln: Aulis.

Maltz, D. N., & Borker, R. A. (1982). A cultural approach to male-female miscommunication. In J. Gumperz (Ed.), *Language and social identity* (pp. 197–217). Cambridge: Cambridge University Press.

Mead, G. H. (1934). *Mind, self and society.* Chicago: University of Chicago Press.

Menacher, P. (1994). Erklärungsansätze für geschlechtsspezifische Interessens- und Leistungsunterschiede in Mathematik, Naturwissenschaften und Technik. *Zentralblatt für Didaktik der Mathematik, 26*(1), 1–11.

Meyer, M. R., & Fennema, E. (1988). Girls, boys, and bathematics. In T. R. Post (Ed.), *Teaching mathematics in grades K-8: Research-based methods* (pp. 406–425). Newton, MA: Allyn & Bacon.

Moi, T. (1989). Patriarchal thought and the drive for knowledge. In T. Brennan (Ed.), *Between feminism and psychoanalysis* (pp. 189–205). London/New York: Routledge.

Parsons, J. E., Adler, T. F., Futterman, R., Goff, S. B., Kaczala, C. M., Meece, J. L., & Midgley, C. (1983). Expectancies, values, and academic behaviors. In J.. T. Spence (Ed.), *Perspectives on achievement and achievement motivations.* San Francisco: Freeman.

Parsons, J. E., Adler, T. F., & Kaczala, C. M. (1982). Socialization of achievement attitudes and beliefs: Parental influences. In *Child Development, 53,* 310–321.

Sacks, H. (1974). An analysis of the course of a joke's telling in conversation. In R. Bauman & J. Sherzer (Eds.), *Explorations in the ethnography of speaking* (pp. 337–353). New York/London: Cambridge University Press.

Saxe, G. B. (1990). Culture and cognitve development: Studies in mathematical understanding. Hillsdale, NJ: Lawrence Erlbaum.

Schor, N. (1989). This essentialism that is not one: Coming to grips with Irigaray. *Differences, 2*(1), 38–58

Schütz, A. (1971). *Gesammelte Aufsätze.* Band I. Den Haag: Nijhoff.

Stallings, J. (1985). School, classroom, and home influences on women's decisions to enroll in advanced mathematics courses. In S. F. Chipman, L. R. Brush, & D. M. Wilson (Eds.), *Women and mathematics: Balancing the equation* (pp. 199–224). Hillsdale, NJ: Lawrence Erlbaum.

Steinbring, H. (1991). Mathematics in teaching processes: The disparity between teacher and student knowledge. *Recherches en didactique des mathématiques, 11*(1), 65–108.

Tannen, D. (1990). *You just don't understand. Women and men in conversation.* New York: W. Morrow & Company.

Thürmer-Rohr, C. (1992). *Vagabundinnen. Feministische Essays.* Berlin: Orlanda Frauenverlag.

Ulich, D. (1976). *Pädagogische Interaktion. Theorien erzieherischen Handelns und sozialen Lernens.* Weinheim und Basel: Beltz.

Voigt, J. (1984). *Interaktionsmuster und Routinen im Mathematikunterricht.* Weinheim und Basel: Beltz.

Voigt, J. (1989). Social functions of routines and consequences for subject matter learning. *International Journal of Educational Research, 13,* 647–656.

Voigt, J. (1994). Negotiation of mathematical meaning and learning mathematics. *Educational Studies in Mathematics, 26,* 275–298.

Walkerdine, V. (1989). *The girls and mathermatics unit: Counting girls out.* London: Virago Press.

Wilson, T. (1981). Theorien der Interaktion und Modelle soziologischer Erklärung. In Arbeitsgruppe Bielefelder Soziologen (Eds.), *Alltagswissen, Interaktion und gesellschaftliche Wirklichkeit* (pp. 54–80). Opladen: Westdeutscher Verlag.

APPENDIX: RULES OF TRANSCRIPTION

,	very short pause (max. 1 sec)
(a sec p)	pause lasting "a" seconds
.	lowering the voice
-	maintaining the pitch

	raising the voice
exact	emphasizing
exact	drawling
(one?)	inarticulate, supposed wording

A: let us begin
                                indent means: B partly speaks simultanously with A
B: but we want

A: let us begin
                                indent means: B interrupts A
B: but we want

SHARLEEN D. FORBES

CURRICULUM AND ASSESSMENT:
HITTING GIRLS TWICE?

Some of the discussion in this chapter is about the average female. If she exists at all, the female who is average in all respects is likely to be comparatively lonely. We need to remember when making comparisons between the average female and the average male that, for each different variable being considered, there are overlapping male and female distributions.

As demonstrated by Mary Gray (1994), this means that all of the males and females in the shaded region of Figure 1 can in some sense be regarded as "matched pairs." It is the different tails of the distributions which impact on, and are reflected in, the mean (average) gender difference. For example, the Assessment of Performance Unit (APU) in the United Kingdom (cited in Girls and Mathematics, 1986) found that the top 30 percent of achievers accounted for most of the gender differences in performance in their investigation of 11- and 15-year-old students. In line with a number of overseas studies (for example, Hyde, Fennema, & Lamon, 1990), New Zealand researchers (Blithe, Clark, & Forbes, 1993; Morton et al., 1993) have also found that males either dominate the highest grades or dominate both the highest and lowest grades (giving male and female mark distributions as illustrated in Figure 1).

Whether we are developing teaching methods, curricula, or assessment procedures in mathematics, if these are primarily targeted to students in Group 2 above then they may disproportionately disadvantage females who dominate Group 1 (and vice versa). It is unlikely any single technique will suit all groups of learners. Indeed, women themselves cannot be classified as one homogeneous group. Women from different socio-economic backgrounds, different cultural and ethnic groups bring to their mathematics different prior knowledge, different attitudes, and different expectations.

Gila Hanna's (1988) study of performance of students in a number of countries on a similar mathematics test showed gender differences varied between countries. Maria Tregalos, when talking about Mexican women, stated (1993) that their position in mathematics relative to males was more extreme according to social class. In New Zealand studies, the performance of Maori (indigenous New Zealanders) in mathematics examinations has been markedly lower than that of New Zealanders of non-Maori (generally European) descent (Manatu Maori, 1991). Although Maori often have low socio-economic status, when this factor is controlled for differences in performance remain which may be attributed to cultural factors (Forbes,

G. Hanna (ed.), Towards Gender Equity in Mathematics Education, 71–91.
© 1996 *Kluwer Academic Publishers. Printed in the Netherlands.*

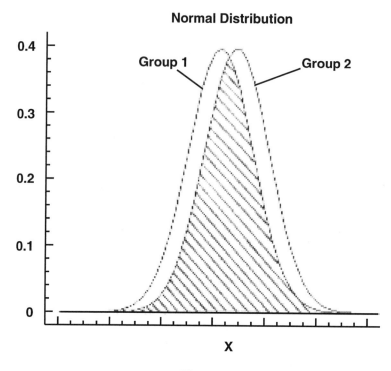

Figure 1

Blithe, Clark, & Robinson, 1990; Garden, 1984). When compared to almost all other groups (except Pacific Island students), Maori girls have the lowest achievement in mathematics (Forbes, 1992; Forbes & Mako, 1993). We need to be aware that many women have a number of factors of disadvantage impacting on their mathematics learning.

Particularly in the formative years when we are aiming not only to impart those mathematics skills needed as a base for life-long learning but also to inspire, motivate, have fun, and develop a love of mathematics, in my view it is essential that a variety of teaching activities, a broad inclusive curriculum, and a mix of assessment methods be used. Unfortunately this has not often been the case in the past. This chapter looks particularly at the one single assessment procedure which has dominated mathematics learning, the timed written test, and compares it with other procedures which may favor females more. Curriculum issues are also raised, and it is conjectured that at the interface between curriculum content and assessment these combine to doubly disadvantage females.

THE NEW ZEALAND SITUATION

New Zealand Senior Secondary School Mathematics

Although overseas research is also used, mathematics in the final three years of the New Zealand secondary school — Forms 5, 6, and 7 (roughly equivalent to Grades 10, 11, and 12 or students with mean ages of 15, 16, and 17 years respectively) — is used to illustrate a number of points in this chapter. Much of the cited research, therefore, is from New Zealand authors.

In New Zealand, mathematics becomes an optional subject for the first time in Form 5 although many schools "encourage" all their students to take it until the end of this year at least. There are a number of "optional" stair-cased national examinations for secondary school students. Students are charged a fee to enter these examinations, which may influence the participation rates of some lower socio-economic groups. The primary purpose of these national examinations is to award qualifications to individual students. As in many other countries however, the results are also used as a barrier to further progress (for example, into restricted-entry tertiary courses) or, on occasions, misused to evaluate either the performance of teachers or of schools.

The international debate about the appropriateness of some assessment procedures has, in part, been stimulated by the differential performance of male and female students in traditional examinations. In New Zealand there have been a number of analyses of secondary school mathematics examinations (Blithe, Clark, & Forbes, 1993; Forbes, 1988, 1994; Forbes, Blithe, Clark, & Robinson, 1990; Forbes & Mako, 1993; Morton, Pemberton, Reilly, Reilly, & Lee, 1988, 1989; Morton, Reilly, Robinson, & Forbes, 1994; Reilly, Reilly, Pemberton, & Lee, 1987; Stewart, 1987). These results all show consistent (but not always significant) gender differences in favor of males. Typically the top grades are dominated by males. However, the studies also indicate that there is a greater range of achievement within each gender than between the genders.

The first national examination met in New Zealand, School Certificate, is at the end of Form 5 (Grade 10) and consists of a set of usually between four to six papers chosen from a wide range of subjects. English is taken by over 80 percent of all school candidates; mathematics is the second most popular subject for both students of European and Maori descent — taken by 75 percent of all male school candidates and 71 percent of all females, but by only 64 percent of Maori male candidates and 59 percent of Maori female candidates in 1991 (Ministry of Education, 1992b). Between 1986 and 1992, students received one of seven grades (D, C2, C1, B2, B1, A2, A1 in ascending order) depending on the marks (out of 100) obtained in the examination. In all these years a larger percentage of males have attained high (A1 or A2) grades than female students. There is a similar gender difference for both students of European descent and Maori students, but

Maori students in general fare worse than non-Maori. The gap in proportions of high-scoring students between those of European descent and those of Maori descent is much more marked than the gender differences (Forbes, 1994; Forbes & Mako, 1993). The most common grade received by Maori students is a C1.

The national examination at Form 6 (Grade 11) consists of a set of internally assessed papers for which grades (from 9 = lowest to 1 = highest) are awarded by individual schools. The grades which a school is given by the national qualifications body (New Zealand Qualifications Authority) to allocate to individual students are, however, determined by that school's overall performance in the previous year's School Certificate examination. The Sixth Form Certificate is thus *moderated* by the School Certificate. Put another way, this year's internal assessment is constrained by last year's written examination. Mathematics is still the second most popular subject across all groups at this level. The participation rate in mathematics, however, has declined considerably for all groups — 58 percent of all Sixth Form Certificate males, 46 percent of all females and 40 percent of Maori males, 30 percent of Maori females in 1991 (Ministry of Education, 1992b). Whereas in School Certificate, a greater proportion of males than females attain the top four grades, in Sixth Form Certificate a greater proportion of females than males attain the top four grades. It is only in the very highest grade in Sixth Form Certificate that males do better than females (Blithe, Clark, & Forbes, 1993).

In the final year of secondary schooling, Form 7 (Grade 12), four to six University Bursary papers may be taken. In addition, up until 1990, students could also choose to sit papers for the prestigious Scholarships examination which was based on the same syllabus as the Bursary. Since 1985, two mathematics papers have been offered: *Mathematics with Calculus* (primarily designed for students intending to do tertiary mathematics) and *Mathematics with Statistics* (designed for students needing mathematics in other disciplines or as a terminating course). Although analyses of national mathematics papers at this level (Forbes, Blithe, Clark, & Robinson, 1990; Morton, Reilly, Robinson, & Forbes, 1993; Stewart, 1981) have shown small overall gender differences in mean performance in favor of males, these differences are not consistent across school authorities (state, integrated [mainly Catholic], or private) or school type (single sex or co-educational). For both males and females, students in single-sex schools perform better than those in co-educational schools (Burgess, 1994; Forbes, Blithe, Clark & Robinson, 1990).

Amount of Mathematics Studied

Although there is some overlap between the two Bursary mathematics prescriptions, students may take both papers. Not surprisingly, the students taking both papers perform better on average than those taking one paper only. This is the case for both genders. However, a smaller percentage of females (38% in 1988) than boys (55% in 1988) take both papers (Forbes, Blithe, Clark, & Robinson, 1990; Morton, Reilly, Robinson, & Forbes, 1994). As stated by Fennema (1979) with respect to an American study:

> A population of males who had spent more time studying mathematics has been compared to a population of females who had studied less mathematics. Since the single most important influence on learning mathematics is studying mathematics, it would indeed be strange if males did not score higher on mathematics achievement tests than did females. (pp. 390–391)

Inter-Curricula Issues

Do the "other" subjects currently taken by females disadvantage them in terms of their mathematics performance? Geraldine McDonald's (1992) explanation for the gender differences in New Zealand's national secondary school examinations between 1986 and 1988 is that the differences result from the statistical manipulation of students' raw scores. A procedure called hierarchical means analysis was used to rank School Certificate subjects. A student attempting any subject was given a place in the ranking of all the students who sat the examination for that subject. The subject itself was also ranked, relative to all the other subjects. Therefore, the higher a subject was ranked, the more students passed that subject. Traditionally male subjects — such as Chemistry, Physics, and Mathematics — were near the top of the subject hierarchy. Sixth Form Certificate is moderated by the school's previous performance in School Certificate so "the gender bias will be carried like pollen on the backs of the grades allocated to the school" (McDonald, 1992, p. 113).

As a result of public pressure, the scaling system in School Certificate has now been dropped, but the practice of moderating Sixth Form Certificate has not.

Some subject combinations (possibly because of overlaps in content) have higher correlations of between-subject student performance than others. In 1989 the New Zealand Universities Entrance Board, with respect to Scholarships examinations, stated that "the large majority of scholarships are gained by students who enter two of their three scholarship papers in Chemistry, Mathematics with Calculus, Mathematics with Statistics, or Physics."

In his analysis of Bursary mathematics performance Burgess found that "combination of subjects is the most influential factor on performance in mathematics" (1994, p. 3). There are marked differences in the subject combinations of males and females at this level. Table 1 gives the ten most popular subjects for male and female Bursary candidates in 1991 (Ministry of Education, 1992b).

Table 1: The ten most popular subject choices in
1991 Bursary examination

Males:		Females:	
Subject:	Percentage of candidates:	Subject:	Percentage of candidates:
Maths with Statistics	63%	English	77%
English	62%	Maths with Statistics	49%
Maths with Calculus	47%	Biology	41%
Physics	38%	Geography	34%
Economics	36%	Maths with Calculus	31%
Geography	33%	History	30%
Chemistry	30%	Economics	28%
Accounting	24%	Chemistry	21%
History	23%	Classical Studies	20%
Classical Studies	21%	Accounting	20%
Total No. Candidates	10627	Total No. Candidates	10983

For males, *Mathematics with Statistics* was the most popular subject choice
for Bursary, and *Mathematics with Calculus* was their third choice. *Mathematics with Statistics*, however, was the second choice for females, and
Mathematics with Calculus the fifth choice.

Participation in Mathematics

Participation in mathematics is a function of past achievement in mathematics, but the relationship is different for males and females. In New Zealand,
those female students who both sat School Certificate mathematics and
then continued into the Sixth Form were less likely to take mathematics
than males *regardless* of the School Certificate grade they had obtained
(Blithe, Clark, & Forbes, 1993; Forbes, 1994). For individual students, the
ranking of their mathematics achievement compared to that in other subjects must also influence decisions on whether to go on. Burgess (1994)
compared School Certificate mathematics marks to other subject marks for
students sitting five or more subjects. When their subjects were ranked by
the mark attained, for most (more than two-thirds) male students, mathematics was one of their top three subjects. This was not the case for
females, for most (more than two-thirds) of whom mathematics was their
third, fourth, or fifth subject.

The issue of different participation rates of males and females in mathematics needs careful consideration. Would females be advantaged in the long term by "filling up" their final year of secondary schooling with mathematics papers and thereby limiting their choice of other subjects? Most Western countries are currently becoming very change-oriented, acknowledging that many people will be required to have several substantially different careers in their working life-time. Is more mathematics the best preparation for this increased demand for flexibility and innovation? Would a broad curriculum, with mathematics integrated in a wide range of subject areas, serve students better?

MATHEMATICS — TOOL OF EMPOWERMENT

It is clear that, at present, mathematics has almost universally taken over the role of Latin as the subject of empowerment. Mathematics is a subject "which filters people out of opportunities in later life" (Mayo, 1994, p. 292). Dale Spender (1986) suggests that the current dominance in many countries of males in mathematics achievement is related to the high status of mathematics. The lack of willingness to accept changes in mathematics which have been advocated on the basis of gender issues supports the theory that mathematics is being used as a filtering device.

When promoting curriculum and assessment change, we must take care not to further disempower women by downgrading the status of mathematics as a result of these changes. Addressing gender inequities in mathematics will contribute to but cannot of itself eliminate the gender inequities in society.

New Zealand research investigating the mathematics skills used in the workplace (Knight, Arnold, Carter, Kelly, & Thornley, 1992) indicated that females had a higher computer usage level than males (with PC-based wordprocessing now replacing typewriters) but used less algebra (symbols to represent numbers) and were less likely to use statistics in decision making. However, in New Zealand Bursary examinations, statistics has been an area in which females have both performed relatively well and shown a preference for in terms of question choice (Forbes, Blithe, Clark, & Robinson, 1990). The differences in mathematics usage were merely a reflection of the type of job in which women were dominant (clerical) compared to men (professional).

Almost a decade ago in New Zealand, the growth in numbers of senior secondary students doing Economics and Accounting was commented on (Walsh, 1987). And Clark and Vere-Jones have noted that:

> the figures indicate an increasing competition between disciplines for the relatively limited numbers of mathematically able students at the upper end of the school programme. (1987, p. 21)

Although slow, this growth has continued. One could speculate that Economics may well become society's next subject of empowerment.

CURRICULUM AND ASSESSMENT

Curriculum

Questions about the nature of the mathematics curriculum are increasingly being asked:

> Is mathematics a set of eternal truths, existing in a world of its own? If it is so pure, so free of values, how come it has such massive effects on peoples' lives? (Tyrell, Brown, Ellis, Fox, Kinley, & Reilly, 1994, p. 337)

Ethnomathematics and the mathematics of indigenous peoples are beginning to be seriously investigated. It is becoming evident that there is no single mathematics curriculum.

In both the Asian-Pacific region and in Europe there are differences between countries in the ages at which children attend school and the ages at which they meet a compulsory mathematics curriculum (Burton, 1993; Clark, 1994a). There are further differences in the topics covered within national mathematics curriculum documents and in the ways in which the topics are delivered. For example, in the Asian-Pacific region, of 12 countries surveyed only the United States and the province of Quebec in Canada taught the various topics separately (Clark, 1994a).

Focussing on the New Zealand situation, Andy Begg (1994) conjectures that there are possibly nine different mathematics curricula. These are:

- the national curriculum
- regional curricula (based on local schemes or informal clusters of schools)
- the school curriculum (school teaching scheme)
- the planned curriculum (an individual teacher's teaching plan)
- the taught curriculum (what is actually taught)
- the learned curriculum (what is actually learnt by individual students)
- the assessed curriculum (what is actually tested)
- the commercial curriculum (textbooks, etc.)
- the hidden curriculum (unplanned learning which influences both social and mathematics education)

All levels of curricula need to be examined for gender and other biases. The excellent attempts by groups such as Family Math (Stenmark, Thomson, & Cossey, 1986) and EQUALS (1988) to address girls' needs at intermediate levels (that is, in the teaching and learning of mathematics) will be limited in their effectiveness unless they are accompanied by similar changes in national and assessed curricula. This need has been recognized by these groups themselves.

In many countries, the gender-specific nature of mathematics curricula has been a topic of debate for a number of years. Concerns that have been expressed include:

- that the subject areas selected either for inclusion or for emphasis in the curriculum reflect "male" priorities (Nancy Shelley [1992] has pointed out that the emphasis and content of many Calculus courses, particularly at the tertiary level, reflect military needs)

- that the topics, or mathematical processes, that more males than females succeed in get elevated in status (Clark, 1993; Kimball, 1989)

- that the language, exemplars, and illustrations in curriculum documents, guidenotes, and textbooks reflect the male world (Northam, 1986).

Curriculum change is a relatively slow and incremental process, usually driven by people other than the teachers of mathematics. The mathematics curriculum in particular, however, is currently under considerable pressure from the rapid changes occurring in technology. Curriculum development in mathematics takes place in different ways in different countries. In France curriculum change is "top-down," academically driven with a centralized national curriculum, but in many other countries it is based on a relativist concept (on children's interaction with their mathematics learning) with emphasis on mathematics for every-day life (Burton, 1993). It is rarely, if ever, undertaken at a national level in order to address equity issues or to increase the participation or achievement of a particular group, such as girls, in mathematics.

New Zealand's current national mathematics curriculum, introduced in 1993 (Ministry of Education, 1992a), was developed by a team of teachers (led by a male) and extends from new primary school entrants to the end of secondary schooling. The curriculum contains five strands: number, measurement, geometry, algebra, and statistics, all nested in an over-arching mathematical-processes framework. The intention is that the emphasis should be on process before knowledge and on the importance of grounding students' learning in their own world. Three components of mathematical processes are identified: problem solving, developing logic and reasoning, and communicating mathematical ideas.

Problem solving has become the current trend in mathematics education not just in New Zealand but internationally. However, many studies have shown males are advantaged by items which require problem-solving skills or an application of mathematics (Armstrong [1985]; Dolittle & Cleary [1987] — both cited in Wilder & Powell, 1989; Hyde, Fennema, & Lamon, 1990) whereas females do well in areas such as computation or algebra (Chipman [1981] cited in Willis, 1990; McPeek & Wilde [1987]; Wily [1986] — both cited in Wilder & Powell, 1989). Meredith Kimball (1993) suggests that this is possibly the reason that arithmetic is considered less important than problem solving. A New Zealand teacher of accelerated mathematics-learning promotes the introduction of formal algebra as soon as children develop arithmetic skills (Tan, 1993). Is the idea destined to be ignored just because girls perform well in algebra relative to boys?

The Curriculum-Assessment Interface

Although the curriculum cannot be seen independently from the assessment processes associated with it, curriculum change is not always accompanied by assessment change. In her study of Asian-Pacific (APEC) countries Megan Clark (1994a) notes:

> The widespread use of multiple choice test items — a "known gender discriminator" — points to a conflict between an absolutist view of mathematics and the clear desire of most APEC members for a curriculum inclusive of girls. (p. 11)

We expect course content to influence the way learning of that content is to be tested, and in general, levels of mathematics achievement are established and defined through the curriculum. We do not, however, expect the testing process to influence the content. This has quite clearly been the case in a number of instances in New Zealand. The School Certificate mathematics paper has been for a number of years a three-hour written examination consisting of short questions answered directly onto the examination script. This led to the testing and teaching of the following "correspondence" associated with the Normal Distribution:

likely = 68% probability = 1 standard deviation

very probably = 95% probability = 2 standard deviations

almost certainly = 99% probability = 3 standard deviations

Not only is this a "nonsense" correspondence but it also has a negative impact on students' future learning of statistics. Two-dimensional vectors and 2x2 matrices remained in this Form 5 (Grade 10) syllabus long after vectors and matrices had disappeared from the two following years of secondary schooling. Presumably, these topics stayed because they suited the examination format. This may be exactly the sort of emphasis on abstraction that Leonne Burton (1986) suggests has made mathematics a masculine discipline. All students "prefer to see the connection between the maths they do and the lives they live" (Purser & Wily, 1992).

Assessment procedures not only affect the mathematics content in the curriculum but also the emphasis given to different sections of the curriculum:

> If some aspect of the curriculum is not examined or given little weighting in assessment then teachers and students may undervalue it. (Begg, 1994, p. 196)

It should be noted that if we are to assess to a norm then we will also teach to that norm.

Curriculum is about context as well as content. Mary Barnes (1992, p. 7, cited in Tyrell, Brown, Ellis, Fox, Kinley, & Reilly, 1994) suggests that the context in which Calculus content is taught is changing:

It is described today as the mathematics of growth and change, and its uses are to be found in nearly every branch of science. We can use Calculus to describe and predict the size of populations, and the effect of changes in the food supply ... and many other issues of vital concern to the life of our planet and the welfare of the people in it. (p. 339)

As female students tend to choose subjects they perceive as socially valuable (Fox, 1977), changes in context such as that described above may improve the participation of women in mathematics.

Context is, however, not solely a teaching issue. It is also an assessment issue. Female students demonstrate a preference for test items set in a real-world context. In a controlled study of New Zealand university students in a first-year statistics-service course, Megan Clark (1994b) showed that when students were given the option of answering one of two questions with identical mathematics *content* but different *contexts* (strengths of concrete or psychologists' tests) they usually answered the first option available. However more female (50%) than male (32%) students picked the second option if the first was about concrete. Overall females (73%) had a clear preference for the psychologist option.

Context can affect performance as well as question preference. There is a body of evidence showing that females perform better on test items which have a "feminine" or gender-neutral context (Clark, 1993; Rosser, 1989; Wilder & Powell, 1989).

As assessment and curriculum are inextricably linked, an understanding of the interface between the two is critical in any analysis of students' abilities. We need to determine the mathematics skills measured by particular assessment methods and to evaluate whether these are in fact the skills we desire our future users of mathematics to have.

Assessment

We need to know not only what our assessment measures but also how good a measure it is, and whether it is a fair measure for all groups. Ideally, all assessment procedures should be both valid (unbiased) and reliable (have low variability) measures of mathematics ability. That is, differences in performance should not just be a product of the testing method itself.

There are a number of forms of assessment in current use in mathematics. These include:

- timed/speed tests (where a set of problems needs to be solved within a short period of time)
- multiple-choice tests (where students chose from either yes/no, true/false, or a range of options)
- essays
- individual projects (usually done over a period of days or weeks)
- group projects

- mastery learning (where basic facts at one level have to be "mastered" before progress to the next level)
- two or three hour written examinations
- oral tests
- "take-home" or "open-book" tests
- internal assessment (generally done by teachers throughout the course and often involving a combination of some or all of the above methods)

Gender differences in mathematics performance are not constant across different forms of assessment. Many of the differences in favor of males noted in the international literature have been in timed written tests. Meredith Kimball (1993) suggests that if classroom grades were to be used instead of standardized examinations to measure mathematics performance then gender differences would be in favor of females. Jan De Lange (1987) noted that male and female students perform much the same on both oral mathematics tasks and "take-home" tasks.

Not all the methods above are discussed in detail in this chapter but it is becoming clear that some types of assessment may unfairly advantage one group of students over another. Only a limited amount of research has been done investigating methods which may best suit women or ethnic minorities. There is a need for more research, particularly controlled experiments comparing one method with another, to determine which assessment methods disadvantage female students least.

The Traditional Mathematics Examination

One method, timed written examinations, has dominated mathematics assessment, particularly at the higher levels. For decades, possibly centuries, there has been a belief in the objectivity of these tests. Not only do many academic mathematicians (who have themselves succeeded under this form of assessment) have faith in this assessment framework but so also do the majority of the general public (including those who have been failed by it). It is becoming clear that these examinations are not necessarily fair. In 1991, the Education Subcommittee of the New Zealand Statistical Association suggested that end-of-course examinations may be neither valid nor reliable measures of ability (cited in Clark, 1994b). Even with the support of professional societies such as this, it may still be extremely difficult to get innovative assessment procedures accepted by society. In many countries, there appears to be far greater reaction to assessment change than to curriculum change.

As Carr (1994) points out even in primary school mathematics in New Zealand:

> The testing movement has always had a presence.... Progressive Achievement Tests in mathematics came to be used widely in schools. Many tasks that most students could do were carefully deleted to ensure that the overall results fitted the statistical "normal curve." (pp. 203–204)

Fortunately, the undesirable effects of these sorts of tests on both teachers and students have been recognized at the primary level, and more innovative methods such as student interviews, portfolios of work, observations, student self-assessment, and co-operative assessment are currently being tried (Carr, 1994). However, there has been little change in national examinations at the secondary level. The most radical change in New Zealand secondary school national qualifications — the replacing of the examined University Entrance in Form 6 by the internally assessed Sixth Form Certificate in 1986 — was not trusted to deliver equitable results to students. Thus, we have the practice of moderating Sixth Form Certificate internal-assessment grades by a school's previous year's performance in the examined School Certificate.

Within the timed-written-test structure there are a number of different areas which have been shown to differentially impact on the performance of females. These include:

Method of test administration: The method of test administration can affect levels of performance and gender differences. Bourke and Stacey (1988 cited in Willis, 1990) showed that achievement was higher overall in a six-problem mathematics test when the items were presented individually than when they were given as one single test, and that the improvement in marks was greater for females.

Question choice: Mathematics examinations often offer students a choice of questions, particularly in national examinations when students from a number of schools are being tested. The intention is to improve the "fairness" of the examination. But male and female students make different choices. In New Zealand, Forbes, Blithe, Clark, and Robinson (1990) and Morton, Reilly, Robinson, and Forbes (1994) found significant gender differences in question choices in Bursary and Scholarship mathematics papers. In Mathematics with Statistics, females opt more for statistics than general mathematics questions; in Mathematics with Calculus, they prefer complex numbers and differentiation questions (both first introduced in the previous year). The preference shown by female students for statistics questions in the Mathematics with Statistics examination may in part be explained by the fact that fewer of them take both mathematics papers. The overlap between the papers is entirely in the general mathematics questions. Also at the tertiary level in New Zealand, gender differences in question choices within statistics courses have been found, with young women preferring and having most success in questions set in either "people" or "the environment" contexts (Blithe, Clark, & Forbes, 1993).

Exam layout: In the past, little attention has been paid to the order in which individual questions, or sections of questions, appear in examination papers. As this does appear to influence the order in which students answer questions, we can no longer dismiss it as unimportant.

In New Zealand, many teachers suggest to students that when they have a choice it is preferable to choose an "easy" question to do first, in order to settle themselves down. A correlation analysis of standardized marks in the Bursary Mathematics with Statistics paper (Forbes, 1988) provided some support for this strategy. Students showed a slight tendency to do better on questions they answered early in the examination (with no apparent gender differences in this tendency).

For a number of years, this paper has consisted of two distinct sections: General Mathematics first, then Statistics second. In general, for both males and females, the mean mark is higher in the statistics section than in the mathematics section. The difference in mean marks between the two sections is, however, greater for females. In spite of the fact that they generally do better on the statistics questions, at least a quarter of all candidates answered their selected questions in exactly the order they appeared in the paper (that is, all the general mathematics questions followed by all the statistics questions). This was only slightly less marked for females than males even though a higher percentage of females chose statistics questions to answer (Forbes, Blithe, Clark, & Robinson, 1990). One could speculate that the order of the sections itself may have unfairly disadvantaged females in this examination. How are choices — such as which set of questions to put first in an examination — made? In this instance, although the author herself was one of the Examiners over this period (setting the statistics questions), the Chief Examiner was male. He determined the examination layout and, although his choice was probably subconscious, it is another example of the decision-making power of the dominant male.

Even within a single question, layout may be an important consideration. In the New Zealand situation again, Clark (1994b) reported that when students in a first-year university statistics examination were given a table containing male and female data with the male data in the left column, female students were significantly more likely to "reach across" the table and compute the female measures first.

Multiple-choice tests: A number of studies in different countries indicate that males are favored by multiple-choice questions in mathematics (Bolger & Kellaghan, 1990; Wilder & Powell, 1989).

Essay questions: Essay questions seem to be little used in mathematics tests. However, it appears that females do perform better when presented with essay-type questions (Eccles & Jacobs [1986]; Slack & Porter [1980] — both cited in Willis, 1990; Burrage [1991] cited in Assessment and Gender, 1992). In the 1987 and 1988 New Zealand Bursary mathematics papers there were only two questions where the gender difference in means was significantly in favor of females. One was an essay question (Forbes, Blithe, Clark, & Robinson, 1990; Morton, Reilly, Robinson, & Forbes,

1994). Essay-type questions provide an opportunity to elaborate on an issue and consider a range of possibilities, but this type of response is all too often limited by the time restrictions imposed on students in testing situations. However, if we are trying to measure the ability of our students to communicate in mathematics, perhaps this is a question form that we should make more use of.

Context: As question context (discussed previously) can affect both the choices and performances of students it must be given careful consideration when designing mathematics examinations. Purser & Wily (1992) illustrated "how much the students, particularly girls, were influenced by the gender orientation of the context of the question."

Comparison of Assessment methods
Time restrictions: There is evidence that eliminating time restrictions in tests can improve the performance of female students (Rosser, 1989).

Internal assessment versus written examination: In New Zealand, there have been two studies comparing mathematics performance in written examinations with that attained in internal assessment. The first compared the performance of individual students in both School Certificate (in terms of their grades: D, C2, C1, B2, B1, A2, A1 in ascending order) and Sixth Form Certificate mathematics (grades 9, 8, 7, 6, 5, 4, 3, 2, 1 in ascending order). In order to analyse the value-added component of the sixth form year, the Sixth Form Certificate male and female grade distributions (for those students who had sat School Certificate in the previous year) were compared within each School Certificate grade. For all School Certificate grades above C1 (about 45%) females were found to do significantly better than males. Although overall a higher percentage of males attain the highest grade (1) this is because fewer females than males achieve high School Certificate grades, and within each School Certificate grade fewer proceed to sixth form mathematics. In Sixth Form Certificate, therefore, females are taking a bigger share of this internally assessed cake than would be expected from their previous written examination performance (Blithe, Clark, & Forbes, 1993; Forbes, 1994).

The second investigation compared the two methods on essentially the same subject matter. The Bursary Mathematics with Statistics paper consists of 80 percent end-of-year three hour written examination, and 20 percent internal assessment (format left to individual schools). Although the internal assessment does vary between schools, in general it is based on one or more practical projects.

After the internal assessment marks are submitted by each school to the central agency responsible for the awarding of qualifications (New Zealand Qualifications Authority), they are *moderated* (scaled to have the same mean and standard deviation) by that school's performance in the written examination. Again, the moderation of internal assessment by examination indicates a lack of trust in either the "fairness" of internal assessment as a procedure, or in the "fairness" of individual schools.

Forbes and Mako (1993) found that not only was the gender difference in the mean overall (final) mark significant (at the 5% level), but so was that for each contributing component. However, while the gender difference in mean examination marks is in favor of males, that for the mean internal assessment marks was in favor of females. It is because the examination mark dominates the final mark by 4:1 that the overall gender difference is in favor of males. Male and female students in state co-educational schools are in similar teaching environments. For these students, the difference in their ranking within their school on each assessment method was also analysed. On average, female students were ranked higher in their school on the internal assessment marks than on the examination marks. The reverse was the case for male students (Blithe, Clark, & Forbes, 1993). The emphasis in Bursary *Mathematics with Statistics* is on the teaching of statistics. Many statisticians hold the view that it is the research, problem-solving, and report-writing skills evaluated in the internal assessment that are the skills needed by practising statisticians, not the ability to regurgitate or manipulate formulae in a set time.

There is, therefore, evidence that the two methods, internal assessment and written examination, each advantage different groups of students in mathematics. It may be that the assessment method itself is preferred by a particular group, or that each method evaluates different skills, or a combination of these. In the Bursary *Mathematics with Statistics* paper however, it would be fairer if the internal assessment and examination were given equal weight. What we are witnessing instead in New Zealand is a move to replace the current internal assessment in this paper with a Common Assessment Task. It is to be hoped that this is not a reaction to reports on the relative success of females in internal assessment. A similar "backwards" development has recently been noted in England by Leone Burton (1993).

Is the practice of *moderation* of internal assessment by examination performance wide-spread internationally? There seems little justification for the moderation of one assessment procedure by another when each method tests different skills.

The Results of Assessment

Developing "fair" assessment procedures will help reduce the current gender inequities in mathematics. However, we also need to ensure that the assessment results are interpreted wisely, whether by the students themselves, their teachers, parents, or by the education administrators.

Attitudes to Success

Males and females view success on our measures of mathematics performance differently. Females habitually attribute their success to hard work or luck and their failures to lack of ability, whereas males see it the other way around (failure is attributed to lack of work or bad luck, and success to ability). In facing advanced work, female students have a decreased expectation of success (Barnes, Plaister, & Thomas, 1987; Howson & Mellin-Olsen, 1986; Leder, 1984; Stipek, 1984). Little research has yet been done to determine whether there is any relationship between the testing method and attitudes to success. In the meantime, we will need to help our female students to become aware of their abilities and to gain greater confidence.

Use of Student Assessment as a Monitoring Tool

A trend which is of concern is the use of results in national examinations as a measure not only of a student's performance but also of a teacher's or of a school's performance. Although there may be some truth in the saying that "a bad doctor loses only one patient, but a bad teacher loses hundreds" we must be extremely careful not to use inappropriate measures to evaluate teacher performance. We know only a little of the interaction between "good" and "bad" teachers and students' performance on tests. When students' marks are used to assess teachers or schools this is usually done with scant regard for different intakes of students or for the "value-added" component (that is, the change in a student's knowledge or skills over the period of study).

As an example, in a New Zealand study of the performance of Form 3 students (mean age 13 years) at the beginning and end of the year on similar mathematics tests, one of the lowest-scoring classes in both tests was found to have one of the highest gains during the year (Forbes, Blithe, Clark, & Robinson, 1990). This class was in a small rural school, all the students were Maori, and the majority were girls. It is likely that the critical factor in the class improvement was the young, enthusiastic Maori teacher. In this particular case, if the students' end-of-year scores had been used as an indicator of his performance as a teacher, he would have received an unfair and incorrect assessment.

SUMMARY

There are mathematics skills which are far more likely to be valued by employers than a specific ability to solve a cubic equation in two minutes flat. The challenge is to find assessment procedures which fairly test the skills we are trying to measure.

Although further research is needed, there is already a body of international evidence supporting the hypothesis that different assessment procedures in mathematics advantage different groups of students. In particular, female mathematics students perform better on internal assessment than on written examinations, when no time restrictions are removed, on essay questions, and on questions which are gender-neutral or set in *their* real world context.

In general, it cannot be appropriate to have assessment procedures consistently designed by one minority section of the population (in the case of mathematics, usually by middle-class academic males) applied to the whole population. It must be fairer and more informative to use a variety of methods to assess a student's attainment, rather than to rely on a single two- or three-hour performance to indicate the outcome of, in some cases, several years of study.

Curriculum and assessment are inter-related. To be effective, changes in one must be accompanied by changes in the other.

The current reality of mathematics for many women is still that of 'hitting' a seemingly immovable barrier. It is likely that the participation, achievement, and — perhaps more importantly — enjoyment of females in mathematics (particularly at higher levels) will increase only with increased female "ownership" of mathematics.

REFERENCES

Assessment and gender. (1992). Wellington: Girls' and Women's Section, Policy Division, Ministry of Education. Background paper for the National Workshop on Professional Development in Educational Assessment, Educational Assessment Secretariat, Ministry of Education, Wellington, New Zealand.

Barnes, M. Plaister, R., & Thomas, A. (1987). *Girls count in mathematics and science.* Perth: Education Department of Western Australia.

Begg, A. (1994). The mathematics curriculum. In J. Neyland (Ed.), *Mathematics education: A handbook for teachers* (Vol. 1). Wellington: Wellington College of Education.

Blithe, T. Clark, M. Forbes, S. (1993). *The testing of girls in mathematics.* Wellington: Author. Report prepared for Women's Suffrage Year.

Bolger, N., & Kellaghan, T. (1990). Methods of measurement and gender differences in scholastic achievement. *Journal of Educational Measurement, 27*(2).

Burgess, D. (1994). *Aspects of bursary mathematics participation and performance, Interim Report.* Wellington: Victoria University.

Burton, L (1986). *Girls into mathematics can go.* London: Holt, Rinehart & Winston.

Burton, L (1993). *Differential performance in assessment in mathematics at the end of compulsory schooling.* Paper presented to the symposium of "International comparisons and national evaluation methods," Fifth Conference of the European Association for Research on Learning and Instruction, Aix-en-Provence, France.

Carr, K. (1994). Assessment and evaluation in primary mathematics. In J. Neyland (Ed.), *Mathematics education: A handbook for teachers* (Vol. 1). Wellington: Wellington College of Education.

Chipman, S. (1988). *Word problems: Where test bias creeps in.* Paper presented at the

Annual Meeting of the American Educational Research Association, New Orleans, USA.

Clark, M. (1993). Friend or foe? *New Zealand Journal of Mathematics, 22*(1), 29–42.

Clark, M. (1994a). *Curriculum development and achievement standards in mathematics education in the asia-pacific region.* Wellington: APEC Human Resources Development Working Group, Ministry of Education.

Clark, M. (1994b). *Assessment issues in the teaching of statistics.* Paper presented to the Fourth International Conference on Statistics Education (ICOTS4), Marrakech, Morocco.

Clark, M., & Vere-Jones, D. (1987). Science education in New Zealand: Present facts and future problems *Royal Society of New Zealand Misc. Series.*

de Lange, J. (1987). *Mathematics: Insight and meaning.* Utrecht: OW & OC.

EQUALS. (1989). *Assessment alternatives in mathematics.* Berkeley: Lawrence Hall of Science.

Fennema, E. (1979). Women and girls in mathematics — Equity in mathematics education. *Educational Studies in Mathematics, 10,* 389–401.

Forbes, S. D. (1988). Mathematics with statistics scholarships examination. *The New Zealand Statistician, 23*(1), 17–25.

Forbes, S. (1992). *Age 13: Mathematics and me for the New Zealand Maori girl.* Paper presented to the Seventh International Conference on Mathematics Education (ICME7), Quebec, Canada.

Forbes, S. (1994). *Assessing statistics learning,* Paper presented to the Fourth International Conference on Statistics Education (ICOTS4), Marrakech, Morocco.

Forbes, S., Blithe,T., Clark, M., & Robinson, E. (1990). *Mathematics for all? Summary of a study of participation, performance, gender and ethnic differences in mathematics.* Wellington: Ministry of Education, New Zealand.

Forbes, S., & Mako, C. (1993). *Assessment in education: A diagnostic tool or a barrier to progress?* Paper presented to the 1993 World Indigenous Peoples Conference on Education, Wollongong, Australia.

Fox, L. (1977). *The effects of sex role socialization on mathematics participation and achievement.* Baltimore: Women and Mathematics: Research Perspectives for Change, Papers in Education and Work, No 8 John Hopkins University.

Garden, R. (1984). *Mathematics and ethnicity.* Wellington: Department of Education, New Zealand.

Girls and mathematics. (1986). London: Joint Mathematical Education Committee of The Royal Society and The Institute of Mathematics and Its Applications, London.

Gray, M. (1994). *The citizen's duty to understand and the statistician's duty to ensure understanding: Statistics in a litigious society.* Paper presented to the Fourth International Conference on Teaching Statistics (ICOTS4), Marrakech, Morocco.

Hanna, G. (1988). *Girls and boys about equal in mathematics achievement in eighth grade results from twenty countries.* Toronto: The Ontario Institute for Studies in Education. Paper presented at the Sixth International Congress on Mathematical Education, Budapest, Hungary.

Howson, A., & Mellin-Olsen, S. (1986). Social norms and external evaluation. In B. Christiansen, A. Howson, & M. Otte (Eds.), *Perspectives on mathematics education.* Holland: Publisher?

Hyde, J., Fennema, E., & Lamon, S. (1990). Gender differences in mathematics performance: A meta-analysis. *Psychological Bulletin, 107*(2), 139–155.

Kimball, M. (1989). A new perspective on women's math achievement. *Psychological Bulletin, 105*, 198–214.

Kimball, M. (1993). Panel contribution. International Commission on Mathematics Instruction, "Gender and Mathematics," Hoor, Sweden.

Knight, G., Arnold, G., Carter, M., Kelly, P., & Thornley, G. (1992). *The mathematical needs of New Zealand school leavers*. Palmerston North: Massey University.

Leder, G. (1984). Sex differences in attributions of success and failure. *Psychological Reports, 54*(1), 57–68.

Manatu Maori. (1991). *E tipu, e rea. Maori education — current status*. Wellington: Ministry of Maori Affairs, New Zealand.

Mayo, E. (1994). On power and equity: Let's not blame the victim. In J. Neyland (Ed.), *Mathematics education: A handbook for teachers* (Vol. 1), Wellington: Wellington College of Education.

McDonald, G. (1992). Are girls smarter than boys? In S. Middleton & A. Jones (Eds.), *Women and education in Aotearoa II*. Wellington: Bridget Williams Books.

Ministry of Education. (1992a). *Mathematics in the New Zealand curriculum*. Wellington: Author.

Ministry of Education. (1992b). *1991 New Zealand Senior School Awards and Examinations*. Wellington: Demographic and Statistical Analysis Unit, Ministry of Education, New Zealand.

Morton, M., Pemberton, J., Reilly, B., Reilly, I., & Lee, A. (1988). Is the gender gap in Bursaries Mathematics with Calculus narrowing? *The New Zealand Mathematics Magazine, 25*(4), 3–21.

Morton, M., Pemberton, J., Reilly, B., Reilly, I., & Lee, A. (1989). Bursaries Mathematics with Calculus revisited. *The New Zealand Mathematics Magazine, 26*(3), 17–40.

Morton, M., Reilly, B., Robinson, E., & Forbes, S. (1993). A comparative study of two nationwide examinations: Maths with Calculus and Maths with Statistics. *Educational Studies in Mathematics, 26*, 367–387.

Northam, J. (1986). Girls and boys in primary maths books. In L. Burton (Ed.), *Girls into maths can go*. London: Holt, Rinehart & Winston.

Purser, P. M., & Wily, H. M. (1992). *What turns them on: An investigation of motivation in the responses of students to mathematics questions*. Christchurch, NZ: Burnside High School.

Reilly, B., Reilly, I., Pemberton, J., & Lee, A. (1987). Gender differences in performance in Bursaries Pure Mathematics. *The New Zealand Mathematics Magazine, 24*(1), 19–23.

Rosser, P. (1989). *The SAT gender gap: identifying the causes*. Washington, DC: Centre for Policy Studies.

Shelley, N. (1992). *Mathematics: Beyond good and evil?* Paper presented to the Seventh International Congress on Mathematical Education (ICME7), Quebec, Canada.

Spender, D. (1986). Some thoughts on the power of mathematics. In L. Burton (Ed.), *Girls into maths can go*. London: Holt, Rinehart & Winston.

Stenmark, J., Thomson, V., & Cossey, R. (1986) *Family math*. Berkeley: Lawrence Hall of Science.

Stipek, D. (1984). Sex differences in children's attributions for success and failure on math and spelling tests. *Sex roles, 83*(3).

Stewart, C. (1987). *Sex difference in mathematics.* Wellington: Research Report Series No 4, Department of Education, New Zealand.

Tan, C. (1993). *My good luck with teaching mathematics.* Paper presented to the New Zealand Association of Mathematics Teachers Conference "Mathematics with Class," Christchurch, New Zealand.

Tregalos, M. (1993). Workshop contribution. International Commission on Mathematics Instruction, "Gender and Mathematics," Höör, Sweden.

Tyrell, J., Brown, C., Ellis, J., Fox, R., Kinley, R., & Reilly, B. (1994). Gender and mathematics: Equal access to successful learning. In J. Neyland (Ed.), *Mathematics education: A handbook for teachers* (Vol. 1). Wellington: Wellington College of Education.

Walsh, B. (1987). *CERTECH: The secondary-tertiary student flow 1979–1986; and the implications for a changing economy.* Wellington: Central Institute of Technology.

Wilder, G., & Powell, K. (1989). *Sex differences in test performance: A survey of the literature.* New York: College Board Report Number 89-3, College Entrance Examination Board.

Willis, S. (1990). *'Real girls don't do maths': gender and the construction of privilege.* Victoria: Deakin University Press.

ANN HIBNER KOBLITZ

MATHEMATICS AND GENDER: SOME CROSS-CULTURAL OBSERVATIONS

According to folklore, the eminent mathematician Hermann Weyl occasionally declaimed: "There have been only two women in the history of mathematics. One of them wasn't a mathematician [Sofia Kovalevskaia], while the other wasn't a woman [Emmy Noether]." Most people in the mathematical community today would not venture so far as to endorse this pronouncement. Yet an astonishing number of mathematicians, mathematics educators, and historians of mathematics subscribe to the belief that there have been only a handful of women in the history of mathematics and that women's numbers in mathematics-related fields today are uniformly low everywhere in the world. Moreover, many people would agree with some form of the statement that the female nature and mathematical thought are incompatible. This paper challenges these beliefs and argues that the historical legacy and current status of women in mathematics are complicated and often contradictory. In the interests of brevity, I have organized what follows as a series of statements.

STATEMENT 1:
The historical legacy of women in mathematics has been mixed.

There have been women mathematicians since classical times. During at least the second half of the 19th century as well as the whole of the 20th century, many women have been active participants in and contributors to the mathematical community (Grinstein & Campbell, 1987). Seven women are generally cited in superficial overviews (Coolidge, 1951; Mozans, 1974; Osen, 1974): (1) Hypatia (370?–415? AD), who according to legend lived in Alexandria, did work on conic sections, and was martyred by Christians; (2) Emilie du Châtelet (1706–1749), courtier and *philosophe*, who translated Newton into French and extensively commented on his work; (3) Maria Gaetana Agnesi (1718–1799), of "witch" of Agnesi fame, who occasionally lectured at the University of Bologna and turned down an appointment to the Academy of Sciences to devote herself to a religious life; (4) Sophie Germain (1776–1831), who proved an important case of Fermat's Last Theorem and whose work on elasticity won her the Grand Prix of the French Academy of Sciences in 1816; (5) Mary Fairfax Somerville (1780–1872), an English polymath and popularizer whose *On the Connexion of Mathe-*

G. Hanna (ed.), Towards Gender Equity in Mathematics Education, 93–109.

matics to the Physical Sciences went through numerous editions in the 19th century; (6) Sofia Kovalevskaia (1850–1891), the first woman to receive a doctorate in mathematics (in the modern sense of the term) and winner of the Prix Bordin of the French Academy of Sciences; and (7) Emmy Noether (1882–1935), one of the founders of modern algebra.

These women are the most famous. But it would not be difficult to compose a list of a hundred or so prominent women mathematicians from many time periods and diverse cultures.

Women mathematicians have often been the ones to break down educational barriers and open up professional opportunities for all women. Some examples:

- The first woman in modern times to be fully integrated into professional, academic life at the university level in Europe was a mathematician. Sofia Kovalevskaia joined the faculty of Stockholm University in 1884 and became an ordinary (full) professor there in 1889. Kovalevskaia was also the first woman to be elected corresponding member of the Russian Imperial Academy of Sciences; the rules were changed to permit her membership (Koblitz, 1993).

- The first PhD granted to a woman in any field by Columbia University (and the first doctorate in mathematics given to any woman by a US university) was awarded to a mathematician, Winifred Haring Edgerton, in 1886 (Green & LaDuke, 1990).

- In 1920, Emmy Noether became the first woman in any field to obtain full qualifications to teach in German universities (Junginger, 1993).

- The first Nigerian woman to obtain a doctorate in any field was a mathematician; Grace Alele Williams received her PhD from the University of Chicago in 1963 (American Association for the Advancement of Science, 1993).

- The first woman to be awarded a full professorship in a scientific or technical field in Vietnam (where full professor is very rare as a title) was a mathematician, Hoang Xuan Sinh. Sinh, a former student of A. Grothendieck and L. Schwarz, has several times been the only female professor involved in the International Mathematics Olympiads (in her capacity as a coach of the Vietnamese team).[1]

Moreover, there are a couple of instances in which the attitude of male mathematicians toward their female colleagues has been particularly impressive:

- In 1896, a survey of the German professoriate was taken on the question of whether women should be admitted to universities with the same rights as men. Mathematicians were unanimously in favor, physicists only slightly less so. Historians, however, were almost all opposed to the entry of women (Koblitz, 1993).

- Carl Friedrich Gauss, whom many mathematicians consider the greatest number theorist who ever lived, attempted to arrange for Sophie Germain to obtain a doctorate from Göttingen University. Gauss had tremendous admiration for what Germain had achieved in mathematics. He wrote:

> But when a person of the sex which, according to our customs and prejudices, must encounter infinitely more difficulties than men to familiarize herself with these thorny researches, succeeds nevertheless in surmounting these obstacles and penetrating the most obscure parts of them, then without doubt she must have the noblest courage, quite extraordinary talents, and a superior genius. (Quoted in Edwards, 1977, p. 61)

On the other hand, one can chronicle enough cases of blatantly discriminatory conduct of male mathematicians towards their female colleagues to remove any doubt that male mathematicians are as much a product of their culture as any other occupational group. Some examples:

- Sophie Germain was a mathematical correspondent of Gauss, Lagrange, and others, and, as mentioned above, her work on the elasticity of metals won the Grand Prix of the French Academy of Sciences. Nevertheless, her name was not on the original list of prize winners on the Eiffel Tower, even though her work contributed to making the tower itself possible (Mozans, 1974, p. 156).

- Christine Ladd-Franklin, a student of J. J. Sylvester at Johns Hopkins University, completed all work for the PhD, up to and including her soon-to-be published dissertation, in 1882. The university did not award her a degree at the time, and in fact did not do so until 44 years later, in 1926. Ladd-Franklin's work earned her one of the very few stars given to women in early editions of *American Men* [sic] *of Science* (Green & LaDuke, 1990, p. 123).

- D. E. Smith and J. Ginsburg note in their *A History of Mathematics in America Before 1900* that *Mathematische Annalen* published 15 articles by US mathematicians in the period 1893 to 1897. They list 14 authors, all male. The *only* one they omit is the one woman — Mary Frances Winston, a student of Felix Klein at Göttingen (Green & LaDuke, 1990, p. 117).

- Emmy Noether was an editor of *Mathematische Annalen* in the late teens and twenties, as the whole of the European mathematical community knew well. She was not listed on the masthead of the journal (in contrast to Sofia Kovalevskaia, who 30 years earlier was listed as editor of *Acta Mathematica*). Moreover, she was sometimes disrespectfully referred to by her colleagues as "Der Noether."

- In the course of several years of giving talks in mathematics departments on Sofia Kovalevskaia, I have sometimes encountered mathematicians who assume demeaning stereotypes about her. They say, for example, that she must have been Weierstrass's mistress, or that the French mathematicians gave her the Prix Bordin out of gallantry, or (my personal favorite) "no, no, *no*, dear, you've got it all wrong. Kovalevskaia was an *amateur* mathematician; her *husband* was responsible for the Cauchy-Kovalevskaia Theorem." (That particular piece of nastiness has several layers. There *was* a male mathematician named Kovalevskii, slightly younger than Kovalevskaia and not in her field; the Cauchy-Kovalevskaia Theorem is sometimes called Cauchy-Kovalevskii, due to differing transliterations of her name; her husband was also a well-known scientist — a paleontologist!)

Sometimes, people have a tendency to get rather smug and complacent when reading about examples of historical discrimination against women or about the situation in countries far away. They congratulate themselves that they are not as stupid as the Johns Hopkins professors were when they refused Christine Ladd-Franklin her degree, or as whoever it was who kept Emmy Noether off the masthead of Mathematische Annalen. It is well to remember, though, that just because certain formal prohibitions and rules barring women have been abolished in academia today, that does not mean that discrimination against women in mathematical fields no longer exists. It behooves us to be especially alert to contemporary methods of discrimination as well as their historical analogs. Although discriminatory practices are in no sense intrinsic to mathematics, nor necessarily worse in mathematics than in other fields, nevertheless there are many ways in which women mathematicians can be the victims of unfairness.

It seems obvious, for example, that a person who can cavalierly and ignorantly propagate sexist myths about historical women mathematicians (such as Hermann Weyl's insults or the nonsense about Kovalevskaia cited previously) will not be above concocting analogously scurrilous stories about present or prospective female colleagues. And that, of course, could be quite harmful. I shall return to this point in Statement 8.

STATEMENT 2:

There is immense variability in the position of women in mathematics both historically and cross-culturally. One cannot necessarily predict what the status of women will be even across ostensibly similar cultural and economic settings.

The percentage of women receiving PhDs at US universities, for example, has varied by over 300 percent; I use "varied" as opposed to "increased" because the percentage has by no means increased steadily. Women begin at a small non-zero percentage of PhDs awarded,[2] and constituted 14.3 percent of mathematics doctorates in the period 1900 to 1940. This compares satisfactorily to the 14.1 percent of doctorates received by women overall in the same period. In the 1940s through the 1960s the percentage of women doctorates in mathematics declined severely, as it did in all fields, so that during the period 1950 to 1969 the figure was always under 6 percent. Not until the 1980s did the percentage of US women receiving doctoral degrees climb past the earlier figure of 14.3 percent, and in the late 1980s and early 1990s it is between 18 and 28 percent, depending on whether one counts allied fields like statistics and computer science. No longer is this close to the percentage of women PhD recipients overall, however. That has peaked at approximately 36 percent (Almanac, 1991; Green & LaDuke, 1990; Keith, 1991; National Research Council, 1979). One can see similar up and down trends in other countries. For example, pre-revolutionary Russia to about 1900 had more women receiving doctorates in mathematics than in the next 20 years. The numbers start to rise again in the experimental 1920s (Koblitz, 1988a; Lapidus, 1978).

Although one might assume that the percentages of female mathematicians in countries with similar cultures or with similar economic indices might be comparable, in fact there are many obvious (and not so obvious) counter-examples. Among the obvious are Great Britain and northern Europe in general as opposed to the US and southern and eastern Europe. The US, France, Italy, Portugal, Turkey, and Spain have significantly higher proportions of female mathematicians than England, western Germany, Sweden, and so on (Burton, 1990; Lovegrove & Segal, 1991; Ruivo, 1987; Stolte-Heiskanen, 1991). Nor can we assume that level of material advancement necessarily correlates positively with percentage of women in mathematics. Japan's and Singapore's numbers are relatively low, while Mexico's, China's, Brazil's, Cuba's, and Costa Rica's are higher (Azevêdo et al., 1989; Burton, 1990; Faruqui, Hassan, & Sandri, 1991; Graf & Gomez, 1990; Koblitz, 1988b; Ruivo, 1987).

Analogously, neighboring countries with similar cultures can also have different percentages of women in mathematical areas. Costa Rica's national university has 12.5 percent women on its mathematics faculty, including a woman chair. Nicaragua's national university, by contrast, has virtually no female professors in mathematics or the physical sciences (though the statistics department at the León branch of the university is chaired by a woman).

In so-called developing countries women appear to be leaders in their departments no less often — and frequently more often — than in the US and Canada. In preparing this paper, I came up with a list of women who were or had recently been heads of mathematics, statistics, or computer science departments in India (Madras Christian College, Lucknow, Mangalore), the Philippines (University of the Philippines and De La Salle University), the University of Costa Rica, University of Nicaragua-León, Mexico, Ivory Coast, Peru, Hong Kong. This does not necessarily mean that the situation for women in mathematics is better in these countries, merely that certain stereotypes need to be treated with caution. One cannot assume that academic communities in Asia, Africa, and Latin America will be more backward on women's issues than in those of North America and Europe. For example, the Third World Academy of Sciences membership is 15 percent female, as compared to approximately 5 percent for the academies of science of the US and the former USSR (Salam, 1988).

At the student level, too, the picture can be varied and can run counter to some of our stereotypes and the images put forward in the popular press. Women's participation in post-secondary level programs in mathematics and computer sciences in some countries (Liberia, Cuba, Indonesia, South Korea, Kuwait, Saudi Arabia, Turkey, Albania, and Italy) approaches or even exceeds their percentage in all programs at that level (UNESCO, 1993, Table 3-14). Other countries (the US, Canada, and most countries of northern and western Europe) have far lower percentages of women students in mathematics and computer science (Chipman, Brush, & Wilson, 1985; Stolte-Heiskanen, 1991).

STATEMENT 3:

Periods of social pressure and reaction can have special consequences for women in "non-traditional" fields like mathematics.

These effects can be either positive or negative (or mixed). In fact, sometimes "good things" can happen for unusual, or even for the wrong, reasons. Take, for example, the situation in Mexico. Like many countries all over the world, Mexico is experiencing economic difficulties. The national universities are in fiscal crisis, and (disproportionately male) professors are leaving in droves for the private universities and the business sector. At a roundtable discussion of women mathematicians and scientists in Mexico City in 1991,[3] the question was raised of the impact of the crisis on university women, whether it might not be a blessing in disguise that the men are abandoning the university to women. True, the percentage of women on the mathematics faculty is rising. But prestige and salary are falling, and many women have the feeling that the best students are not going to be attracted to mathematics anymore.

This is a complex situation, and one that has analogs in many other countries and fields. If a specialty becomes "saturated" (that is, too many professionals to make good salaries) or unattractive for some other reason, opportunities for women can increase.[4] Also, a field can become stratified. That is, the teaching of mathematics (and university teaching in general) can become more or less the province of women, while the academies of sciences and research institutes can remain largely or entirely male. One sees this in many countries that overall have decent percentages of women in mathematics, such as Mexico, India, China, Nigeria, and Costa Rica. The most prestigious positions and places become or remain largely male preserves.

STATEMENT 4:
One must be very careful about making generalizations from
historical and cross-cultural comparisons.

Mindless use of one or another indicator to make a sweeping generalization about women's status in mathematics can give misleading results. It is not possible to use the same indicators to determine the situation in every country. The significant statistic might be the percentage of women teaching at the university level. But it might also be the proportion of women at research institutes and academies of sciences (and at what level), or the percentage of women who publish (or who publish in foreign as opposed to domestic journals), or the proportion of women who go abroad for conferences, post-graduate study, and so on, or the percentage of women awarded grants by national and international funding agencies. Indices can have different meanings in different countries, and the prestige of various positions and honors can vary considerably. This is not to say that it is unimportant that women constitute a large percentage of professors on a mathematics faculty in a certain country. But this measure might not be the unique indicator of women's success or status in the mathematical world.

STATEMENT 5:
Despite the problems and societal obstacles, women can fare
reasonably well in mathematics for a number of reasons.

There are, after all, some recognized and relatively objective standards in the mathematical community. Women might have to be better than their male counterparts to be judged equal, but the standard is not impossible to achieve. Sofia Kovalevskaia, for example, offered *three* works to Göttingen in fulfillment of her degree requirements. She and her adviser Karl Weierstrass reasoned that as the first woman applying for the doctorate, her case would have to be especially strong. Three *were* sufficient, however.

Moreover, there is perhaps not so great an "Old Boy Network" in mathematics as in other fields and more of a tradition of tolerance and eccentricity. There is also a certain pride in traditions of internationalism — in not wanting to allow political, ethnic, or ideological considerations to interfere with one's mathematical judgments. These factors have worked in women's favor in a large variety of contexts.

Women can also fare well in mathematics because of a relatively small number of men who have served as excellent mentors, in a couple of cases despite otherwise not terribly progressive politics or overall commitment to women's rights. Some random examples:

- Karl Weierstrass performed valuable mentoring functions not only for Sofia Kovalevskaia but also for her friend Iulia Lermontova. It was through Weierstrass's intervention with university officials at Göttingen that Lermontova was able to become the first woman in the world to receive her doctorate in chemistry (Koblitz, 1993).

- Felix Klein supported Mary Winston, Grace Chisholm, and Margaret Maltby in their admission to graduate studies in mathematics at Göttingen University in 1893. He later said that the women were fully comparable to his male students.

- Of the seven PhDs in mathematics awarded to women by Johns Hopkins University before 1940, five were students of one professor (Morley). Eleven of the 13 women who completed PhDs at Catholic University were students of Landry. The University of Chicago granted 46 of the 229 doctorates given to women in the US through 1939; of these, 30 were students of either Leonard Eugene Dickson (18 of his 67 students were women) or Gilbert Ames Bliss (12 of his 52 students were women) (Green & LaDuke, 1990).

- Lee Lorch, now emeritus at York University, performed mentoring functions for many women — including several of the few black women who later received PhDs in mathematics — during the time he taught at Fisk University.

STATEMENT 6:
Cross-cultural disparities in female educational and employment patterns in mathematics and computer science raise serious questions about theories that claim innate gender differences in mathematical ability.

We need to ask ourselves whether there are other explanations: faulty test design, socio-cultural factors, and so on. Theories must be examined critically and tested historically and cross-culturally. Several generalizations have become quite popular in late-20th century US society, yet they rest on dubious foundation and would fail under historical and cross-cultural scrutiny.

For at least three decades the received wisdom — and the line being pushed by "objective scientists" like Camilla Benbow and Julian Stanley — was that girls are better at verbal tests and boys at mathematical ones (Benbow & Stanley, 1980). Now, though, that picture is breaking down in several important ways. The so-called gender gap on US standardized tests narrowed considerably during the 1980s, to the point that specialists at the Educational Testing Services (ETS) in Princeton are now saying the differences are not statistically significant. Also, several recent studies have shown that even in mixed groups where males had performed noticeably better than females on mathematics Scholastic Aptitude Tests (SATs), on other tests, including ETS's Mathematics Achievement Test itself, there were no significant gender differences. (For a reasonably current review of the literature see Kenschaft, 1991b.)

More importantly, the supposed gender differences are constant neither across ethnic groups within the US and Canada nor across cultures. A 1987 study noted that even within the mathematics SAT, the gender gap varies considerably. It is largest for Hispanics and smallest for Afro-Americans (Ruskai, 1991). Moreover, Gila Hanna has pointed out in her comparative studies that differences *between countries* are much larger than those between boys and girls, and that the gender gap is larger in countries with low scores (like the US) than in those with high scores (like Hungary and Japan). On the geometry test, for example, US males scored 39.7 while US females scored 37.9 — a mere 1.8 points. Meanwhile, the advantage over the US of Hungary and Japan was relatively massive, since all subgroups in both countries scored between 55 and 60 points (Hanna, 1989).

Some countries (Thailand and South Korea, for example) do not exhibit any statistically significant differences between male and female performance on mathematics achievement tests (Hanna, 1989; Hanna, Kündiger, & Larouche, 1990; Kwon, in press). As Hanna and her collaborators note, the one clear conclusion that emerges from all the statistics is that one must doubt the biological explanations of male/female difference in mathematical ability, since it is "very unlikely" to vary between countries (Hanna, Kündiger, & Larouche, 1990, p. 96).

Other studies have also pointed to the culture-bound nature of many of our notions of gender difference. For example, spatial ability tests given to Native American children in Alaska and to central African children show either no difference or one favoring the females (Fausto-Sterling, 1985; Kenschaft, 1991a; Koblitz, 1987; Ruskai, 1991). Needless to say, these tests are *not* the ones reported with fanfare in the *New York Times*.

Benbow's and Stanley's studies have immense cultural problems. Gender differences are not consistent across ethnic groups, and even girls who did worse than boys on the Benbow and Stanley test outperform the boys in school. Moreover, before they test, Stanley's research centre apparently sends parents a pamphlet noting that boys do better than girls on the test. Such a message to parents reveals the researchers' blindness to issues of bias-free experimental design (Jackson, 1990; Ruskai, 1991).

Also dubious are generalizations about a fundamental physiological difference which affects the way men and women reason about mathematics or their interests in mathematics. The latest manifestation of this type of pseudo-scientific study is the brain lateralization "research." This has a long, inglorious history dating back to early 19th-century phrenology (Alper, 1985; Bleier, 1984; Durndell, 1991; Fausto-Sterling, 1985).

STATEMENT 7:

There are fundamental problems with anything that smacks of
essentialism; this includes so-called feminist gender-and-science theory.

"Essentialism" encompasses any theory that attributes gender differences to biological, genetic, psychosocial, or other immutable factors. One problem with such theories is that, as indicated above, women's position in mathematics and the natural sciences is constant neither across cultures nor across time periods. What in one country or time might be considered unfeminine might in another country or time not be so. Specifically with regard to the mathematical sciences, there are several countries, including the Philippines, Turkey, Kuwait, and Mexico, for example, in which women constitute a rather high percentage of mathematics-related professions, especially when compared with the numbers in northern Europe, Canada, or the United States (Faruqui, Hassan, & Sandri, 1991; Lovegrove & Segal, 1991; Stolte-Heiskanen, 1991; UNESCO, 1993; United Nations Statistical Division, 1992). Yet most feminist theorizing about gender and mathematics assumes that women's participation in these fields is uniformly low.

The theorists also assume that Victorian-era bourgeois stereotypes concerning femininity and gender polarities are uniform across all cultures, classes, and historical periods. This is far from being the case, even within western Europe. In Italy, for instance, the stereotype is different from that in the US or Sweden or the UK. Women are purported to be "natural" theoreticians (hence the relatively large number of Italian women mathematicians and computer scientists), while men are supposed to be more practical by nature (and thus become engineers rather than theoretical scientists).

Moreover, the position of women in mathematics and the sciences can change quite rapidly for the better (or worse). The changes are far too rapid to be explainable by biological theories of difference or by psychosocial theories such as Nancy Chodorow's (1978).[5] Gender and science theorists have the unfortunate tendency to make generalizations despite clear historical and cross-cultural counterexamples. For example, it is simply not true that women's status in the sciences has remained unchanged since the Scientific Revolution, though that is a claim often made by theorists like Evelyn Fox Keller (1985), Sandra Harding (1986, 1991), and their imitators.

In like manner, I am disturbed by certain aspects of work by Belenky and her collaborators (1986), Gilligan (1982), and others on women's purportedly different ways of knowing. Belenky and her collaborators, for example, assert that the gender differences they describe are independent of culture, ethnicity, and class. This is highly questionable. A colleague of mine at Hartwick College, Katherine O'Donnell, has conducted years of field research with poor migrant farm women in New York state. At the beginning she had the expectation of finding support for the writings of Belenky and Gilligan (O'Donnell, in press) Now, however, she is firmly convinced that, on the contrary, the interactions of gender, culture, race, and class are far too complex to be encompassed by a simplistic gender-polarity theory (O'Donnell, forthcoming).

To bring this discussion closer to mathematics, let us take the work of Sherry Turkle (1984) on children and computers. Under the influence of object-relations theory and gender-and-science theory, Turkle has created the concepts of "soft mastery" and "hard mastery" to categorize her analysis of the use of computers by school-age boys and girls. For Turkle, "hard mastery is the imposition of will over the machine through the implementation of a plan ... the hard masters tend to see the world as something to be brought under control." Soft mastery, on the other hand, is more interactive — "the soft masters are more likely to see the world as something they need to accommodate to, something beyond their direct control." Though Turkle gives examples of soft masters of both sexes, she says that girls tend to be soft masters, while hard masters are "overwhelmingly male."

Large parts of the theory sound quite plausible at first. Certainly it is true that in most societies males and females are socialized differently, with female socialization tending more towards valuing qualities like accommodation and interaction, and male socialization tending more towards control. Upon reflection, however, some people have raised questions about the validity of the theory and voiced concerns about its social implications.

It has been pointed out (by Beth Ruskai [1990], for example) that the theory contains certain assumptions about the nature of computer science that are rather far off the mark. What Turkle dichotomizes as "hard" and "soft" mastery might be better characterized as two inextricably interwoven parts of the creative scientific process. The attempt to label people as being *either* hard *or* soft, therefore, misses a crucial point of what it is to do science.

Turkle's theory makes no allowances for historical and cross-cultural variation. It ignores the evidence that women's participation in the sciences — including computer science — has varied widely from one decade to the next and even between neighboring countries (Koblitz, 1991; Lovegrove & Segal, 1991). Moreover, stereotypes regarding women's innate capacities and their relation to mathematics and computer science vary from culture to culture.

There is also reason to be uncomfortable with the way Turkle's theory dovetails with current western European and North American stereotypes about women's intrinsic nature. In practice, these stereotypes can contribute to discrimination against women in the workplace, and to the segregation of women in so-called pink-collar ghettoes like data processing.

Sherry Turkle is fairly skillful in inserting caveats in her generalizations and in reminding us that the picture is complex. (Harding, Belenky, and Gilligan are not nearly so careful.) Unfortunately, however, the caveats and reservations rarely make it into popular accounts of the work. What gets picked up in the media is the rather simplistic idea that girls cannot be attracted into computer-related fields unless the machine can be portrayed as artistic, relational, and "soft." We do not work in a vacuum. If the media *can* distort a theory, they *will*.[6]

In the US, one of the most common manifestations of sexism and racism in the classroom is a refusal to intellectually challenge girls and members of minority groups. They are condescended to and patronized and do not receive adequate exposure to the more rigorous, thought-provoking, and elegant aspects of mathematics. Their understanding thus rarely attains the level of the systematic and the structural; they seldom arrive at the stage where they can see much point in doing mathematics. Unfortunately, this phenomenon can be exacerbated by overzealous followers of the ideas of Turkle, Gilligan, and Belenky.

STATEMENT 8:

Any discrimination against younger faculty automatically falls disproportionately on women because of the demographics of the profession. The injustice is increased because of certain usually unconscious but quite pervasive attitudes about women's "natural" roles.

In the US, for example, women are often assigned heavier teaching loads than men and more courses at the introductory level. This is a triple blow: more work is devoted to teaching as opposed to research; teaching evaluations are automatically worse because of the nature of the course (introductory courses virtually always receive lower evaluations than upper division courses); and any deviation from stereotypically feminine behavior (such as attempting to enforce high academic standards) is met with displeasure (and low ratings) by students. Numerous studies indicate differential treatment of women faculty on evaluations. For example, students expect to be "nurtured" by women and punish them for deviations from the ideal "feminine" standard (N. Koblitz, 1990).

Women faculty are caught in a bind. Either they devote tremendous time to teaching, in which case their research suffers, or they devote as much time to their research as their male colleagues do, remaining aloof from students, in which case they are penalized more heavily than men on evaluations.

Moreover, the evaluatory process for granting tenure is in essence a blackballing system.[7] Even if 80 percent of the department have not a sexist bone in their bodies, the opposition of a relatively few curmudgeons can sink a woman's chances in a variety of ways. It is rather like the old anti-Semitic blackballing system for admission to US country clubs — the system persisted for so long because all that was needed was one person in opposition.

There is an analogous blackballing system in place for hiring and tenure in academic departments today. A couple of people can skew the process and, in fact, wreck it. Consider the following situations:

- Say a woman has children and resumes mathematical activity after a couple of years hiatus. How does one interpret this? One could talk about the fact of her return to active research as indicating resolve and high mathematical ability and dedication. But one typically hears the diehard sexists referring instead to the "unfortunate gap in her publication record."

- Women are often held up to far higher standards than men on the pretext of not lowering standards. For a woman's appointment, there cannot be the least shadow of a doubt, while men are often given the benefit of the doubt.[8] We all know at least one senior professor who thinks nothing of spreading stories to the effect that it would be lowering the department's standards to hire a woman, even when the woman being considered is a far better researcher than he is himself.

In the US a certain amount of sound and fury signifying little or nothing has sprung up around the issue of Affirmative Action. Departments sometimes have women come on campus for interviews, under pressure from the administration, only to have the appointment sabotaged by a couple of diehard sexists working assiduously to undermine the process. There is a lot of rumor-mongering that goes on about Affirmative Action. Traditionalists routinely exaggerate the success of the program, telling female graduate students things like, "oh, *you'll* have no problem getting a job, you're in fashion these days," or jokingly advising their male students to wear a skirt to the job interview. This kind of childish and disingenuous behavior in itself creates a bad atmosphere for women. It conveys the impression that colleagues do not have confidence in the women they have hired and implies that any women in the department are there on sufferance.

CONCLUSION

My goal here clearly has not been the presentation of a definitive account or a polished treatise. It was my intention to throw out food for thought, to illustrate some of the curiosities and ironies inherent in women's position in mathematics historically and across cultures. The interactions of gender and culture are never simple or straightforward. The history of women in mathematics has not been some Whiggish triumphal passage from darkness into light, but neither has it been a chronicle exclusively of discrimination and marginalization. The history and present status of women in mathematics are complicated, and often the picture has elements of contradiction. One conclusion seems obvious, however. The complex and multifaceted interactions of gender and mathematics can be understood only if one takes into account historical and cross-cultural perspectives.

NOTES

1. Much of my information on women mathematicians of Asia, Africa, and Latin America does not come from published sources, which are generally conspicuous by their absence. Rather, the data emerge from extensive personal interviews with the women themselves, their male colleagues, officials in women's organizations and ministries, and so on.

2. Graduate education in the US and Canada did not really get started until the late 1870s, so there were some women involved in the enterprise almost from the beginning.

3. The discussion was part of a week of activities commemorating the centenary of Sofia Kovalevskaia's death.

4. Veterinary medicine in the US, pharmacy in El Salvador, medicine in the USSR and The Philippines are examples of this phenomenon.

5. In general terms, object-relations theory says that a girl infant never has to disidentify herself from her mother. Therefore, she never sees the world as alienated from herself in the same way a boy infant does. A boy baby, on the other hand, realizes very early that he is not the same gender as his mother. He has to distance himself from her and thus begins to objectify the world. Because object-relations theory attributes gender differences in intellectual outlook to an immutable mechanism of early childhood, it is almost indistinguishable from a genetic or biological theory. Moreover, there is no way the theory can account for differences between individuals or for change over time or across cultures.

6. We had a distressing illustration of exactly this point in Swedish newspaper coverage of the 1993 ICMI Study Conference "Gender and Mathematics Education" in Höör, Sweden. The headline of the 9 October 1993 *Dagens Nyheter* story was "Mathematicians disagree whether biology makes a difference," and ICMI President Miguel de Guzmán was misquoted as saying that gender differences manifest themselves from day one in the classroom!

7. I realize that the concept of tenure (the right to relative job security, granted after a probationary period of some years) is not institutionalized in all countries. But I believe that the following discussion is applicable to most other kinds of hiring and promotion processes as well.

8. This kind of discrimination includes behavior such as the infamous cases of analogous curriculum vitae being sent to chairs with male and female names attached — the chairs routinely recommended the women for lower positions than the men! Bernice Sandler has documented many examples of this sort of (unconscious) prejudice in her "Chilly Climate" series (Sandler, n.d.).

REFERENCES

Almanac. (1991). *The chronicle of higher education*, XXXVIII, No. 1.

Alper, J. S. (1985). Sex differences in brain asymmetry: A critical analysis. *Feminist Studies*, **11**(1), 7–37.

American Association for the Advancement of Science [AAAS]. (1993). *Science in Africa: Women leading from strength*. Washington, DC: Author.

Azevêdo, E. S., et al. (1989). A mulher cientista no Brasil. Dados atuais sobre sua presença e contribuição. *Ciência e cultura, 41*(3), 275–283.

Belenky, M., Clinchy, B., Goldberger, N., & Tarule, J. (1986). *Women's ways of knowing: The development of self, voice and mind.* New York: Basic Books.

Benbow, C. P., & Stanley, J. C. (1980). Sex differences in mathematical ability: Fact or artifact? *Science, 210,* 4475, 1262–1264.

Bleier, R. (1984). *Science and gender.* New York: Pergamon.

Burton, L. (Ed.). (1990). *Gender and mathematics: An international perspective.* London: Cassell.

Chipman, S. F., Brush, L. R., & Wilson, D. M. (Eds.). (1985). *Women and mathematics: Balancing the equation.* Hillsdale, NJ: Lawrence Erlbaum Associates.

Chodorow, N. (1978). *The reproduction of mothering: Psychoanalysis and the sociology of gender.* Berkeley: University of California Press.

Coolidge, J. L. (1951). Six female mathematicians. *Scripta Mathematica, 17,* 20–31.

Durndell, A. (1991). Paradox and practice: Gender in computing and engineering in Eastern Europe. In G. Lovegrove & B. Segal (Eds.), *Women into computing.* London: Springer-Verlag.

Edwards, H. M. (1977). *Fermat's last theorem: A genetic introduction to algebraic number theory.* New York: Springer-Verlag.

Faruqui, A. M., Hassan, M.H.A., & Sandri, G. (Eds.). (1991) *The role of women in the development of science and technology in the third world.* Singapore: World Scientific Publishing.

Fausto-Sterling, A. (1985). *Myths of gender: Biological theories about women and men.* New York: Basic Books.

Gilligan, C. (1982). *In a different voice: Psychological theory and women's development.* Cambridge, MA: Harvard University Press.

Graf, H. B., & Gomez, H. G. (1990). Acerca de las científicas en la UNAM. Unpublished paper delivered at the National Autonomous University of Mexico. (I am indebted to the authors for giving me a copy of this paper.)

Green, J., & LaDuke, J. (1990). Contributors to American mathematics. In G. Kass-Simon & P. Farnes (Eds.), *Women of science: Righting the record.* Bloomington: Indiana University Press.

Grinstein, L. S., & Campbell, P. J. (1987). *Women of mathematics: A biobibliographical sourcebook.* New York: Greenwood Press.

Hanna, G. (1989). Mathematics achievement of girls and boys in grade eight: Results from twenty countries. *Educational Studies in Mathematics, 20,* 225–232.

Hanna, G., Kündiger, E., & Larouche, C. (1990). Mathematical achievement of grade 12 girls in fifteen countries. In L. Burton (Ed.), *Gender and mathematics* (pp. 87–97). London: Cassell.

Harding, S. (1986). *The science question in feminism.* Ithaca, NY: Cornell University Press.

Harding, S. (1991). *Whose science? Whose knowledge? Thinking from women's lives.* Ithaca, NY: Cornell University Press.

Jackson, M. V. (1990). SATs ratify white male privilege. Association for Women in Mathematics *Newsletter, 20*(5), 9–10.

Junginger, G. (1993). A woman [sic] career in veterinary medicine. Lecture given at the XIXth International Congress of History of Science, August 1993.

Keith, S. Z. (1991). A statistical overview of American women doctorates, 1988–1989. In P. C. Kenschaft (Ed.), *Winning women into mathematics* (pp. 59–60). Mathematics Association of America.

Keller, E. F. (1985). *Reflections on gender and science*. New Haven: Yale University Press.

Kenschaft, P. C. (1991a). Fifty-five cultural reasons why too few women win at mathematics. In P. C. Kenschaft (Ed.), *Winning women into mathematics* (pp. 11–18). Mathematics Association of America.

Kenschaft, P. C. (Ed.). (1991b). *Winning women into mathematics*. Mathematics Association of America.

Koblitz, A. H. (1987). A historian looks at gender and science. *International Journal of Science Education, 9*(3), 399–407.

Koblitz, A. H. (1988a). Science, women, and the Russian intelligentsia. *Isis, 79*, 208–226.

Koblitz, A. H. (Ed.). (1988b). *La mujer en la ciencia, la tecnología y la medicina*. Seattle: Kovalevskaia Fund.

Koblitz, A. H. (1991). Women in computer science: The 'soft mastery' controversy. *Kovalevskaia fund newsletter, VI*(2), 5–6.

Koblitz, A. H. (1993). *A convergence of lives. Sofia Kovalevskaia: Scientist, writer, revolutionary* (2nd. ed). New Brunswick, NJ: Rutgers University Press.

Koblitz, N. (1990). Are student ratings unfair to women? Association for Women in Mathematics *Newsletter, 20*(5), 17–19.

Kwon, O. (in press). U.S.-Korea cross-national studies on college entrance exams.

Lapidus, G. W. (1978). *Women in soviet society*. Berkeley: University of California Press.

Lovegrove, G., & Segal, B. (Eds.). *Women into computing*. London: Springer-Verlag.

Mozans, H. J. (1974). *Woman in science*. Cambridge, MA: MIT Press. [Original publication date, 1913.]

National Research Council, Commission on Human Resources. (1979). *Climbing the academic ladder: Doctoral women scientists in academe*. Washington, DC: National Academy of Sciences.

O'Donnell, K. (in press). A class act. *Phoebe*.

O'Donnell, K. (forthcoming). *Got nowhere to go and no way to get there: Migrant women's struggle for dignity*.

Osen, L. M. (1974). *Women in mathematics*. Cambridge, MA: MIT Press.

Ruivo, B. (1987). The intellectual labour market in developed and developing countries: Women's representation in scientific research. *International Journal of Science Education, 9*(3), 385–391.

Ruskai, M. B. (1990). Why women are discouraged from studying science. *The Scientist, 4*(5), 17, 19.

Ruskai, M. B. (1991). Are there innate cognitive gender differences? Some comments on the evidence in response to a letter from M. Levin. *American Journal of Physics, 59*(1), 11–14.

Salam, A. (1988). Speech given at the Third World Academy of Sciences, Trieste, October 3.

Sandler, B. R. (n.d.) The campus climate revisited: Chilly for women faculty, administrators, and graduate students. *Project on the status and education of women.* Association of American Colleges.

Stolte-Heiskanen, V. (Ed.) (1991). *Women in science: Token women or gender equality?* Oxford: Berg.

Turkle, S. (1984). *The second self: Computers and the human spirit.* New York: Simon and Schuster.

UNESCO. (1993). *1992 Statistical yearbook.* New York: United Nations.

United Nations Statistical Office. (1992). *Statistical yearbook 1990–1991.* New York: United Nations.

BARBRO GREVHOLM

WOMEN'S PARTICIPATION IN MATHEMATICS EDUCATION IN SWEDEN

The performance and participation of women in mathematics education have been the focus of much research in the area of gender and mathematics education. In Sweden, performance seems not to be a problem. Girls' achievements in mathematics at the end of compulsory school, as measured by marks, are better than those of boys (Grevholm & Nilsson, 1994). In upper secondary school, the most advanced mathematics course is taken in the natural science program, and also here girls have better end-of-school marks than boys (Grevholm & Nilsson, 1994). The same result is found in the social science program. As far as it has been documented, women seem to have as good results as men in undergraduate studies in mathematics. Accordingly, performance will not be discussed here but the focus will be on participation, which on the contrary seems to be a serious problem in Sweden, perhaps worse than in other Western countries.

In most countries in Western Europe, women are a minority in mathematics and mathematics education. Sweden is no exception, although the country often claims to have come far in the development of equality for women and men. In 1993, the Minister of Education appointed a committee called Female and Male in School and gave them the task:

- to collect relevant knowledge and experience about the importance of gender for the development and schooling of children and young people
- to define, based on this knowledge, what equality in connection with schooling means and in this way give the work for equal conditions a real meaning
- to suggest and initiate actions to disseminate knowledge about the importance of gender belonging (Utbildningsdepartementet, 1994)

In 1969, it was decided that the compulsory school system must work towards equality and the same was mandated for upper secondary school in 1970 (Skolöverstyrelsen, 1969, 1970). The actual situation has changed very slowly, however, and in 1985 demands were made of the Ministry of Education that education about gender and equality should be compulsory in all undergraduate teacher education (Statistics Sweden, 1986). It is interesting that this is what the Female and Male in School committee suggested as a result of the investigation conducted in 1994 (Utbildningsdepartementet, 1994). The most important suggestion to come from the committee's work is that equity should be a compulsory field of knowledge in teacher education and should be treated as a pedagogical issue.

G. Hanna (ed.), Towards Gender Equity in Mathematics Education, 111–124.
© 1996 *Kluwer Academic Publishers. Printed in the Netherlands.*

COMPULSORY SCHOOL

Sweden has a compulsory school system that consists of nine school-years. All pupils study mathematics all nine years for three to four hours per week. For the last three years, it was possible (but is no longer) to choose between a general or advanced course in mathematics. In the beginning of the seventies, a few more boys than girls choose the advanced course, but by the end of the eighties this difference had disappeared and about 60 percent of both girls and boys chose the advanced course (Grevholm, 1995a). Why this difference disappeared has not been investigated and there are no obvious reasons other than the tendency for girls to make less gender-stereotyped choices.

UPPER SECONDARY SCHOOL

Since 1993, the upper secondary school has been undergoing a process of change, and the new structure is expected to be in place by 1995. The former 27 different education lines are being changed into 16 programs, where two are theoretical and the others vocational. Nearly all youngsters from a year-group go to upper secondary school for three years. The upper secondary school in Sweden has been a girls' school and a boys' school in the sense that particular educational streams have been dominated by either one sex or the other. One reason for changing the organization has been the desire to break this tendency of traditional career choice, but it is doubtful whether the new system will help to achieve this change.

If a student wants to carry on with university studies in mathematics, science, or technology, the obvious way is to choose the natural science program in upper secondary school, where one studies the most advanced mathematics course (about 300 one-hour lessons). This choice is made by a little less than 20 percent of a year-group and of these about one-third are girls.

The number of students going through the natural science program is not adequate for Sweden's needs for trained scientists and technologists. Therefore, it is also possible after finishing upper secondary school in the social science program to take an extra year to complete the necessary courses for studying science and technology at university. According to prognoses by the education authorities (Berg, 1993), the required percentage of people needed to enter the natural science program is about 30 percent of a year-group, and there is obviously a potential group of girls able to do so successfully. The data on mathematics performance show that girls are successful in this program. They have as good results at national standardized tests as boys and they get as good marks as boys or even a little better (Grevholm & Nilsson, 1994; Ljung, 1990).

Are there tendencies of development? Table 1 shows the percentage of females among all students graduating from the natural science program of upper secondary school (or equivalent education) at different times (Statistics Sweden, 1986, 1994).

Table 1: Percentage of females graduating from the
natural science program

Year	1974	1984	1994
Female students	15%	20%	33%

The participation of female students is growing, but a closer look reveals that there are fluctuations in the percentage of females from year to year. The obvious aim must be to get 50 percent of female students in this program. It should be as natural for females as for males to choose this education. The new curriculum doesn't mention the problem of gender imbalance in connection with mathematics and no special initiatives are suggested. There is only a general statement about working for greater equality between the sexes (Skoverket, 1994a, 1994b).

UNDERGRADUATE EDUCATION

For the academic year 1992/93, the number of first-year university students was 30 percent greater than the year before. Women made up 60 percent of all first-year students. The extended number of places at universities is the result of government decisions regarding the economic situation and problems in the labour market. Of all registered undergraduate students, females make up 57 percent (VHS, 1994). How are these students distributed in different areas of education that include mathematics?

Undergraduate education in Sweden is organized so that a student can either choose a full study program for three to five years aimed at some special career or choose one short course at a time, thus composing studies that have a certain perspective but are specific for the individual. In the first case, the student is guaranteed further studies as long as he or she is successful, but in the second case there is no guarantee of a place in the next course chosen.

TEACHER EDUCATION

Education to become a teacher for the compulsory system in mathematics and science for Years 1 to 7 takes three-and-a-half years, and in 1992 the percentage of women among those students was 78. Teachers for Years 4 to 9 are educated in a certain combination of subjects. To become a teacher of mathematics and one other subject for Years 4 to 9, the program consists of four years. The dominant combination is mathematics and science, and this is chosen by about 80 percent of education students. The number of women varies according to the combination of subjects. All data presented in the tables below have been calculated by the author from figures ordered from the educational database of Statistics Sweden.

Table 2: Teacher students for compulsory school,
years 4–9, percentage of women, 1992

Combination of subjects	%
Arts/mathematics	83
Home economics/mathematics	69
Sports/mathematics	37
Mathematics/science	58
Music/mathematics	59
Textile work/mathematics	100
Wood carving/mathematics	0
Total	56

In the group of students studying to become a teacher of mathematics in compulsory school, women dominate strongly for Years 1 to 7 with 78 percent, and not so strongly with 56 percent for Years 4 to 9. This should be viewed in connection with the fact that being a teacher in Sweden has low status and the status has decreased more and more during the last ten years. Also the salaries for teachers are low compared with other academic groups with the same length of studies and the same demands in the work situation (Statistics Sweden, 1994).

The program to become a teacher in upper secondary school lasts for four-and-a-half years and the student is educated in mathematics combined with some other subject. Reorganization of this level has been carried out since 1987. In 1987, this program was for both teachers in upper secondary school and teachers in compulsory school for Years 7 to 9. Therefore the number of university places referred to is not the same for the two years compared. The proportion of women varies in these combinations as shown in Table 3 below.

Table 3: Teacher-students for upper secondary school
percentage of women enrolled 1987 and 1992

Combination of subjects	1987	1992
Mathematics/computer science	–	20
Mathematics/physics	29	28
Mathematics/chemistry	64	47
Social science/mathematics	38	31
Mathematics/natural science	57	–
Total	41	31

The new organization explains the lower participation of women in 1992 in the sense that more women choose to teach younger children. There are also, nowadays, other ways to become a teacher of upper secondary school. But the number of women in this program remains at such a low level that it is worrying.

Barely one-third of the students in 1992 were women, and this is in contrast with other subject combinations such as languages, arts, and the humanities, where women are a majority. Also in upper secondary school, teachers are becoming more and more a group dominated by women. This group has also been affected by a decrease in status and a relative decrease in salaries. It is obvious that the combination of math with chemistry attracts more women while physics and computer science have little attraction for them. The small figure for the combination with social science should be interpreted by the fact that it is difficult to find a position in a school with this combination. The conclusion is that the group of students in the natural science program which is only one-third female meets a group of teachers in mathematics which is only one-third female. It is clear that in the three groups — Years 1 to 7, 4 to 9, and upper secondary — the proportion of women teaching decreases the older pupils are.

INDEPENDENT COURSES IN MATHEMATICS

One week of full-time studies in a university course corresponds to one point: the first term corresponds to studies at the level 1 to 20 points, and so on. In 1992, the number of students in independent courses was about two-thirds of the number of all teacher-students together (about 2400 to 3500). In this group we find 24 percent of women in total in 1992 (26% in 1987), but they are not equally distributed throughout the levels. Table 4 shows the percentage of women at different levels of study.

Table 4: Independent courses in mathematics,
percentage of women enrolled 1987 and 1992

Study level	1987	1992
1-20 points (first term)	30	28
20-40 points (second term)	14	12
40-60 points	17	15
61+ points	20	7

Here the development over time seems to be negative for women. The number of places for students has increased considerably. There is no obvious explanation for the decrease in participation of women, and it has not been investigated or discussed.

At the first level, the number of women is about the same among the students as in upper secondary school in the natural science program. We could say that girls are as willing as boys to study mathematics at the beginning of university. But then the number of women among the students of mathematics decreases drastically. No efforts have been made to investigate why this is so. But the problem has been discussed lately, and some development programs have started.

MATHEMATICAL STUDY PROGRAMS AT THE UNIVERSITY

In 1992 the number of students registered in any natural science program at the university was over 4000, and about 45 percent of those were women. But if we look at the programs that have mathematics as a major component, the picture is different. Table 5 shows figures from some programs with mathematics.

Table 5: Students in programs with mathematics,
percentage of women 1992

Study program	% of women
Main subject mathematics	33
Mathematics/computer science	19
Mathematics/chemistry	40
Mathematics/statistics	100
Mathematics/physics/computer science	28
Total	26

The dominant combination is mathematics and computer science, and the number of students taking statistics was so small this year that no general conclusions should be drawn from that. Again we see that the combination with chemistry attracts more women.

Has there been a development over time in the participation of women in natural science programs at the universities? It is not possible to make a direct comparison because the structure of the courses has been changed during the years. But it is possible to get a general perception of the situation by comparison with the courses on hand in 1987. Table 6 shows the proportion of women.

Table 6: Students in natural science programs in 1987,
percentage of women enrolled

Study program	%
Biology program	53
Physics program	18
Geology program	40
Chemistry program	59
Mathematics program	25
Total	40

It might be relevant to compare the 1987 total of 40 percent with the total in 1992 of 45 percent. Also it might be possible to compare the 25 percent in the 1987 mathematics program with the 26 percent total in programs with mathematics 1992. Maybe this could be interpreted as an increase of women's participation. The tendency of chemistry, biology and geology to attract more women is obvious. Physics and computer science seem to be most unattractive to women.

The number of students graduating from two programs, the mathematics program in the faculty of natural science and the statistics program in the faculty of social science, could be of interest. Table 7 gives data for a recent ten-year period.

Table 7: Graduates during the years 1983–92, percentage of women

Years	83	84	85	86	87	88	89	90	91	92
Maths	26	26	35	16	23	26	40	30	22	26
Statistics	–	–	–	–	–	71	33	35	47	80

The average value for the period 1983 to 1992 for mathematics is 27 percent, and for the years 1988 to 1992 for statistics it is 50 percent. The result for mathematics corresponds well to the number of female students registered in the program, which may confirm that women are successful in their studies. The difference between the statistics and mathematics programs is remarkable.

The proportions of women in independent courses in mathematics (24%), in mathematical programs (26%), and in teacher education for upper secondary school in mathematics (31%) in 1992 are similar and correspond well with the number of girls in the natural science program in upper secondary school, which is a necessary prerequisite for these university studies. An interesting question is: If the number of girls in the natural science program in upper secondary school were increased, would the proportion of women in the previously mentioned university programs then increase automatically in relation to that? Teachers at the university are aware of the fact that the number of students in the natural science program is the critical point for getting more students to science and technology. What is then the situation in studies of technology?

STUDIES TO BECOME AN ENGINEER

In general, the various engineering programs are very male-dominated but in some areas there has been considerable change during the years. Table 8 shows the participation of women at three different times.

Table 8: Students training to become engineers,
percentage of women 1968, 1979, and 1989

Year	68	79	89
Program			
Architecture	24	45	51
Chemical engineer	13	40	51
Civil engineer	1	17	25
Industrial & management engineer	–	14	23
Mechanical engineer	1	7	14
Engineering physics	4	14	11
Electrical engineering	1	7	10
Computer science and engineering	–	7	7

Table 9: Shorter technical programs,
percentage of women, 1989

Computer engineer	8
Production engineer	8

(Source: Statistics Sweden)

The number of women has increased considerably in architecture and chemistry, but in other areas the development is slow, and the participation of women is lower than in other university areas with mathematics. Worth noting is the fact that when the number of women in chemistry in the technical university in Gothenburg exceeded 54 percent in 1993, people from the university drew to the attention of politicians in parliament the "fact" that this was a threat to the profession. It could cause lower status for the career, lower salaries, and other problems (Member of Parliament M. Edgren, personal communication, 1993). A comparable discussion has not taken place in any case where men are in the majority in a program.

The general problem for technical education is that too few students want to enrol. This of course means that the quality of the students is not as high as desired when they start their education. There is much discussion going on in Sweden concerning these problems, and both professional organizations and the larger society are taking action to stimulate students to go into science and technology (Grevholm, 1993).

POSTGRADUATE STUDIES

Research education in mathematics and related fields can take place in a faculty of mathematics and science or a faculty of technology. Results shown in the following tables include both of these groups. As a comparison, we look at the program of statistics, which belongs to the faculty of social science. Women participate in research education to an even lesser extent than in undergraduate studies in mathematics. Has there been development over the years?

Table 10: Research students registered in some subjects, percentage of women

Year	85	86	87	88	89	90	91	92
Subject								
Mathematics	12	10	14	13	11	11	10	10
Math. stat.	15	17	11	22	26	29	25	29
Num. analys.	16	10	5	8	3	19	13	18
Statistics	22	29	32	24	26	28	27	32

Table 11: Research students registered in some subjects, number of women

Year	85	86	87	88	89	90	91	92
Subject								
Mathematics	7	7	10	11	9	9	10	11
Math. stat.	5	5	4	11	14	15	15	18
Num. analys.	7	5	2	3	1	5	4	7
Statistics	13	18	21	17	21	22	25	30

During the period 1985 to 1992 the average proportion of women registered in mathematics was 11 percent, in mathematical statistics 23 percent, numerical analysis 11 percent, and statistics 28 percent. As the figures fluctuate a lot from year to year it is difficult to see trends. What can be seen is that in mathematics there is no increase in the number of women and that the participation in mathematics and numerical analysis is low compared to the other two subjects. Maybe there is a trend of rising numbers in mathematical statistics. The participation of women in postgraduate studies in mathematics is too small compared with the number of women in undergraduate studies.

To what degree do women graduate from research education? The number of PhDs or doctors of technology are so small that it is interesting to look at a longer period. During the years 1983 to 1992 the proportion of women that received a doctoral degree in mathematics was 6.9 percent, in mathematics statistics 8 percent, in numerical analysis 10.3 percent, and in statistics 19 percent. It is not relevant to compare these figures with the number of registered students, because these are long programs and we would need to have different time periods. It can be said that, if the relation in numbers between the subjects hasn't changed during the years, it is relatively more rare for a woman to get a degree in mathematics and mathematical statistics than in the other two subjects.

To be able to relate these figures to something, we may compare them with the number of registered research students in some other areas. See Tables 12 and 13.

Table 12: Research students registered in some faculties, percentage of women enrolled in 1985 and 1992

	1985	1992	Change during period, % of 1985
Humanities	69	47	-32
Theology	16	26	+63
Law	24	39	+63
Social science	36	40	+11
Medicine	30	40	+33
Math/science	24	30	+25
Technology	13	20	+54

Table 13: Research students registered in some faculties, number of women in 1985 and 1992

	1985	1992
Humanities	894	896
Theology	28	57
Law	42	60
Social science	803	953
Medicine	720	1248
Math/science	476	682
Technology	269	652

With the exception of the humanities, the relative participation of women seems to be increasing in the various faculties. The next slowest rise seems to be in mathematics/science, and as we have seen above, this rise doesn't occur in mathematics, but only in the other subjects. Why does science, and especially mathematics, hold such a small attraction for women? Why is the change in this area so slow?

TEACHERS IN SCHOOLS

In the compulsory school system, there are about 24,000 female teachers of mathematics and about 9300 male teachers — that is, 72 percent of the teachers of mathematics are women. This, of course, is explained by the fact that teachers for the younger pupils are to a great extent women. In upper secondary school, there are about 2800 male teachers and a little more than 1000 female. This means that about 28 percent of the teachers of mathematics in upper secondary school are women, quite the opposite situation from compulsory school. This also reflects the proportion of women seen above in the proportions of graduates from teaching programs.

TEACHERS AT THE UNIVERSITIES

It is well known that the number of female teachers at the universities is low, especially in mathematics and related areas. Is there any change going on? (See Table 14.)

Table 14: Teaching staff at universities in certain groups, number of women and men, 1990 and 1993

Area		Social science		Natural science		Technology	
		Men	Women	Men	Women	Men	Women
Professors	1990	186	14	251	10	486	16
	1993	219	21	272	12	489	15
Senior lecturers	1990	658	185	636	105	679	48
	1993	742	237	712	129	768	54

For social science there has been a small rise in the relative proportion of women professors and senior lecturers, but in natural science and technology no progress has been made. Master's assistant is another position that requires only a first degree and most often means that the work consists of more teaching than for a senior lecturer and little chance of doing research. These are the only positions where the number of women increases noticeably (from about 20% to 30% in science and technology). There is no female professor of mathematics in Sweden, and there has not been since

Sofia Kovalevskaia died in 1891 (Grevholm, 1994). She was the first, and so far has been the only, female professor of mathematics in Sweden. The number of female senior lecturers in mathematics is about 10 to 12 (a rise of 50% since 1993, but very low compared to other subjects) from a total number of about 250 to 300 senior lecturers in mathematics. (It is not possible to get precise numbers from Statistics Sweden, because they don't register positions by subject, only by faculty.)

CONCLUSIONS

The statistical data shown here confirm again a well-known picture of women's unequal participation in mathematics and related fields. They also show that there has been almost no change over time in the situation in mathematics. It is very serious that in women's participation in mathematics education there is no positive development worth mentioning, rather — as shown in some respects — a negative one. The development in other subjects differs from mathematics. Where are the invisible barriers for women? There is a desire on the part of the government to bring about change. Five million SEK have been given to five different projects at the universities for development work to get more female students in mathematics. All the projects intend to promote women's participation in education in mathematics and related areas in, for example, women-only groups, problem-based learning, small-group work, or teaching in a gender-inclusive context.

FEMALE AND MALE IN SCHOOL

As mentioned in the introduction, a report has been presented with suggestions intended to change the situation concerning equity in education (Utbildningsdepartementet, 1994). These are some of the suggestions:

- teacher education should be more overtly aimed at equity
- teacher education should be more appropriate in response to the demands of competency for tomorrow in gender questions
- more time for knowledge about gender differences, similarities, and differences in education must be given in teacher education
- the gender imbalance in teacher education must be changed
- equity must be a compulsory field of knowledge in teacher education
- relevant in-service training must be given to teachers
- headmasters and people in leading positions in schools must work with equity as a pedagogical issue

It remains to be studied what changes these suggestions can bring about.

Some teacher-training programs provide non-compulsory courses about gender equity, and the Women's Forums at the universities provide courses like Gender, Science, and Technology. More resources have been given to women's studies, and, for example, a female visiting professor in mathematics will be invited. The Swedish network Women and Mathematics, started in 1990, is constantly growing and disseminates knowledge and research results in conferences and newsletters (Brandell, 1994; Grevholm, 1992, 1995b). It is astonishing how slow the development is concerning equity compared to the flood of investigations and reports during the years. In 1994 the Swedish Parliament has more than 40 percent women members, and the government has 50 percent women. Perhaps these women in politics can make some effort to bring about a faster development for women's participation in mathematics.

REFERENCES

Berg, B. (Ed). (1993). *Att stimulera intresset för naturvetenskap och teknik. Siffror och erfarenheter*. Stockholm: Verket för högskoleservice.

Brandell, G. (Ed.). (1994). *Kvinnor och matematik. Konferensrapport*. Luleå: Luleå University.

Grevholm, B. (Ed.). (1992). *Kvinnor och matematik. Rapporter om utbildning, nr 1, 1992*. Malmö: Lärarhögskolan i Malmö.

Grevholm, B. (1993). *Naturvetenskap och teknik. Kan forskningsinformation stimulera?* Stockholm: Verket för högskoleservice.

Grevholm, B. (1994). Svenska kvinnor i matematiken. In G. Brandell (Ed.), *Kvinnor och matematik. Konferensrapport*. Luleå: Luleå University.

Grevholm, B. (1995a). Gender and mathematics education in Sweden. In B. Grevholm & G. Hanna (Eds.), *Gender and mathematics education* (pp. 185–195). Lund: Lund University Press.

Grevholm, B. (1995b). A national network of women — Why, how and for what? In P. Rogers & G. Kaiser (Eds.), *Equity in mathematics education*. London: Falmer Press.

Grevholm, B., & Nilsson, M. (1994). Assessing mathematics in Europe. Sweden. In L. Burton (Ed.), *Who counts?* (pp. 241–262). Birmingham: Trentham Books.

Skolöverstyrelsen. (1969). *Läroplan för grundskolan*. Stockholm: Liber Utbildningsförlaget.

Skolöverstyrelsen. (1970). *Läroplan för gymnasiet*. Stockholm: Utbildningsförlaget.

Ljung, G. (1990). *Centrala prov i matematik, åk 3 NT-linjerna*, Stockholm: Primgruppen.

Skolverket. (1994a). *Läroplan för de frivilliga skolformerna*. Stockholm: Liber utbildning.

Skolverket. (1994b). *Läroplan för den obligatoriska skolan*. Stockholm: Liber utbildning.

Statistics Sweden. (1990, 1994). *Årsbok om utbildning* [Yearbook for education]. Stockholm: Statistiska Centralbyrån.

Statistics Sweden. (1986). *Kvinno- och mansvär(l)den* [The World of Women and Men 1986. Equal opportunity in Sweden]. Stockholm: Statistiska Centralbyrån.

Utbildningsdepartementet. (1994). *Vi är alla olika* (DS 1994:98). [We are all different]. Stockholm: Fritzes.

VHS. (1994). *Årsrapport om högre utbildning och forskning.* Stockholm: Verket för högskoleservice.

KARI HAG

GENDER AND MATHEMATICS
EDUCATION IN NORWAY

Norway is a sparsely populated country; its 4.3 million inhabitants spread out over an area of roughly 387,000 square kilometres, which yields a population density of 13 people per square kilometre. A basic principle of Norwegian educational policy is that all children and young people have an equal right to education and training, irrespective of sex, social and cultural background, or aptitude. Moreover, university education is free.

There has, however, been very little focus on "mathematics and gender" by the authorities in Norway. In 1991 the Norwegian Mathematics Council, in concert with Agder College and the Secretariat for Women and Research, called a national meeting on mathematics education and women in mathematics. The conference SÅNN, JA! took place in Kristiansand in the spring of 1992. Following are excerpts from the opening address of Anne-Lise Hilmen, Assistant Director, The Research Council of Norway.

> Something must be done! This is the reason why responsible bodies and institutions have taken the initiative to arrange this conference.... Something must be done about the shortage of women in mathematics and something must be done about the content and quality of the teaching of mathematics at all levels of our educational system.
>
> Why is it important to discuss these matters? A few facts from the Norwegian scene, I hope, will convince those of you who do not find it obvious. I restrict myself to some statistics from the research and development section. Research and development is, as we all know, a most dominant factor in our society, creating knowledge and insight and producing competence, used throughout the whole educational system, in business life, in the public sector, in politics, in shaping our minds, thoughts, reflections — [it] is truly overwhelming.
>
> Natural sciences and technology are heavy fields with great impact on society both directly and indirectly by supplying experts and leaders to important positions outside their educational field. For instance, mathematics is both the foundation and the tools in natural sciences such as physics, chemistry, and biology and in technological fields. It also plays an important part in economics and many social sciences. Mathematics is the gateway and for many pupils, a bottleneck for many educations.
>
> At the universities and regional colleges in Norway, the scientific staff in mathematics in 1989 occupies 15 percent of the total natural science staff. Only 7 percent of the mathematicians, that is 10 persons, are women. This places mathematics in the bottom group together with physics and informatics as the most male-dominated research fields together with technology.
>
> How is the situation among the recruitment personnel? — nowadays all of them shall be heading for a PhD. Thirty-five percent of the recruitment personnel in all research fields in 1989 are women. In natural sciences there are 30 percent and in

G. Hanna (ed.), Towards Gender Equity in Mathematics Education, 125–137.
© *1996 Kluwer Academic Publishers. Printed in the Netherlands.*

mathematics 16 percent women.

The external pressure and the internal, institutional work are not giving rapid results, if any. But if you are not impatient, you might find some comfort in the fact that: In natural sciences, 36 percent of the students in 1989 were females, a rise from 16 percent 11 years earlier (1978). The corresponding numbers for mathematics were: 24 percent for the year 1989 vs. 5 percent for 1978.

Finally, if you combine this statistical picture with the fact that the ordinary school system lacks at least 100 teachers with a master's degree in mathematics, you might think there is no end to the misery.... You're quite right!

So — there is nothing else to do than to roll up our sleeves and say: OK — we define this as the bottom level; we can aim in only one direction: The female participation in mathematics on all levels of society shall increase dramatically during the next 10 to 20 years.... The question is *only*: How?

One answer is obviously to do something with the self-image of *little* girls, for instance by giving them good [role] models. Another answer is to teach them mathematics in a way relevant to their experiences and lives.

Is this gloomy picture about to change, now, two years later? It is hard to say. Some recent statistics are better than the 1989 numbers Hilmen referred to, others remain about the same.

HISTORICAL BACKGROUND

The Norwegian school system can be traced back to the cathedral schools of the Middle Ages.

Elementary Education

The first educational act was passed in 1739, and the first elementary schools for the general public were run by the Church. The schools were established for the purpose of educating the populace in the Christian religion. In addition to religious instruction, reading and, usually, writing and arithmetic were also taught.

During the 19th century, elementary education was gradually expanded. More subjects were taught, and the number of years of schooling was increased. In 1889, a seven-year period of compulsory education was introduced, specifically for all segments of the population.

Traditionally the school systems in urban and rural areas were organized differently as far as gender was concerned. In the cities, for example, single-sex classes were common well into the 1950s. It was not until 1959 that girls and boys were put on an equal footing in urban schools. In rural areas, all students had the same number of lessons in mathematics, while in urban areas, boys had two more mathematics lessons a week than girls. In the textbooks, some of the problems were marked by a little asterisk to indicate that these problems were meant only for the boys to solve. Girls had home economics and nursery subjects when the boys were taught solid geometry.

In 1969, the seven years of compulsory education were extended to nine. In 1974 new national guidelines for compulsory schooling, *M74*, were introduced. These guidelines included, among other new issues, a whole chapter dealing with gender equity. *M74* introduced such innovative teaching methods as co-operation and group work. School-based grades in the first six years were abandoned, and evaluation was done informally. *M74* lead to a drive towards decentralization in education and opened possibilities for developing local curricula. The next set of national curriculum guidelines, *M87*, developed these shifts further but did not propose any advances relating to mathematics and gender.

Recently the parliament has decided to lower the starting-school age from seven to six and to extend compulsory education by one more year, to ten years in total. It remains to be decided when this reform will be implemented. (At present, voluntary preschool education is already available to 90% of all six-year olds.)

Post-Elementary Education

Latinskolene were a direct continuation of the cathedral schools. Reforms were carried out in 1869 and 1896, after which these schools became known as the "upper schools." The system consisted of both lower and upper secondary school, leading to the *examen artium*. At first, the upper school was only for boys, just as the Latin schools had been. But in 1878, there was an addition to the *Act of 1869* granting girls the right to graduate from lower secondary school, albeit with fewer hours of mathematics and more hours of handicraft than boys had. Women were allowed to take *examen artium* in 1882. The upper secondary school was transformed by the *Act of 1896* into a co-educational system, a victory for the women's movement at the time. The law also contains as an explicit provision that women shall be given the same rights as men in regard to appointments as school principals and teachers.

In the last decades, Norwegian post-elementary education has seen a rapid expansion in participation. An important goal for educational policy has been to make upper secondary education available to all. In 1976, vocational training and academic studies were combined in a single system of upper secondary education. A further step in this direction was taken when it was decided that, as of August 1994 (*Reform 94*), all young people should have the right to three years of upper secondary education after completing their compulsory schooling.

Norway's first university was the University of Oslo, founded in 1811. There are now universities in Oslo, Bergen, Trondheim (which encompasses the Norwegian Institute of Technology), and Tromsø. In addition there are six specialized colleges at university level and 26 regional colleges. In 1884 women obtained the right to "submit themselves to University examinations." It was not until 1979 that a PhD program was introduced in Norway. Up to that time doctorates were awarded for independent research. The first Norwegian woman to earn a PhD in mathematics, Mary Ann Elizabeth Stephansen, was awarded her degree at the University of Zürich in 1902, and she later became a docent at the Agricultural College of Norway. Stephansen was a lonesome pioneer: 69 years elapsed before the next Norwegian woman was granted a PhD in mathematics, and this was by an American, not a Norwegian, institution!

FEMALES IN MATHEMATICS — THE CURRENT SITUATION

The education system in Norway has three main levels (a well-developed adult education system runs parallel programs at all three levels), as shown in Figure 1. Women constitute over half of the enrolment in upper secondary school and at universities and colleges. There are now about 160,000 students, which represents an increase of approximately 60 percent during the period from 1988 to 1993.

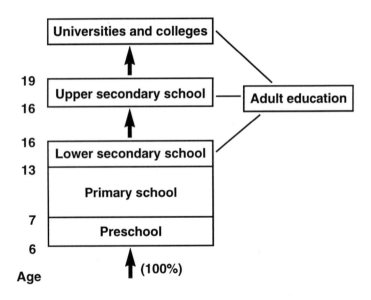

Figure 1

In cases where adequate statistics are available, we will compare certain gender specific demographics in mathematics with those in other subjects. For upper secondary school, sufficient information to give a coherent national picture of the situation for either students or teachers has yet to be amassed.

Primary and Lower Secondary School

Mathematics is obligatory in all grades, and there is only one mathematics curriculum. More than 60 percent of the teachers are women. In lower secondary school, some of the teachers have bachelor of science. degrees from universities (plus pedagogical training), while others have graduated from teacher-training colleges and are certified to teach all subjects in primary and lower secondary school. Until recently, there was no requirement for the latter group to take mathematics at the teacher-training colleges, at which one could enrol with just one year of compulsory mathematics from upper secondary school. (As of 1991, a mathematics unit of study, equivalent to one-fourth of a year, has been required in the program at the teacher-training colleges.) Inquiries among mathematics teachers in primary school have shown that the teachers are well educated in general but not in mathematics: 59 percent of them have no mathematics component in their higher education.

Upper Secondary School

The standard mathematics courses have lately been organized as shown in Figure 2.

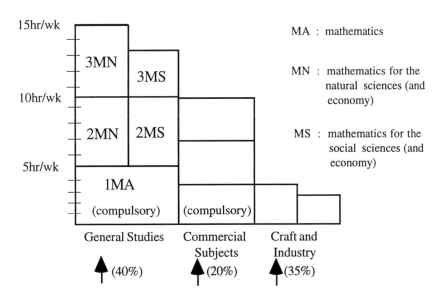

Figure 2

There are no longer special tracks (science, language, etc.) within the area of General Studies. This model, the New Structure, was gradually implemented beginning in the late 1980s.

After the compulsory mathematics course (1MA), whose enrolment consists of roughly 55 percent girls and 45 percent boys, almost half the pupils quit mathematics. *And for every boy quitting there are two girls who do not continue.* Approximately 13 percent of the full age group end up taking further studies in mathematics for the natural sciences (3MN), and 4 to 5 percent take further studies in mathematics for the social sciences (3MS), which has become more popular with the New Structure. According to the latest information, about 35 percent (a drop from 39% in the mid 1980s) of the 3MN pupils are girls, while the percentage of girls among the 3MS pupils is about 50 percent.

Most mathematics teachers in the General Studies courses have master's degrees, but more often in physics than in mathematics. As a rule, these physics majors are males. When Hilmen said the ordinary school system lacked 100 teachers with master's degrees in mathematics, she was referring to the number needed for each upper secondary school to have at least one teacher with a master's degree in mathematics. (Here mathematics also includes mathematics education, which is a new field in Norway.)

Universities

One can study mathematics at all universities in Norway and at many of the colleges. In this discussion, we restrict ourselves to the universities. Here an individual is admitted to a school, not a program, and can take one or more courses in mathematics, a cand. mag. degree (BSc degree) including mathematics, a cand. scient degree (MSc degree), or a dr. scient degree (PhD), which builds upon an master's degree in the subject. To become a mathematics teacher in secondary school, one should have a "20 group" (one-year credit) in both mathematics and one other school subject. A teacher's certificate also requires one year of pedagogical training.

Now to the statistics. The term mathematics is used in a broad sense, as before. Collecting data from Norwegian mathematics departments on students taking a master's degree in recent years, we obtain Table 1.

Table 1: Women in master's programs in
university mathematics departments

| | Mathematics | | Pure math | | Statistics | | Applied | |
	Total	Women	Total	Women	Total	Women	Total	Women
1981-85	170	20	68	11	33	4	69	5
1986-90	168	38	48	7	48	18	72	13
1991-93*	130	38	41	4	46	20	43	14

* (only 3 years)

Thus, the figures confirm a significant increase of late in the number of women. But the increase has come in statistics and applied mathematics. What happens in the stream of pure mathematics? Figures from the University of Oslo (H. Galdal, personal communication, 1994) show that women constitute a stable 29 percent of the students in the standard introductory course, which builds upon 3MN, and in recent years 52 percent of those in the less theoretical, applied course, Mathematical Methods, which assumes only the second-highest level of mathematics from high school. The first course is part of the "20 group" in mathematics, whereas Mathematical Methods is not. The other (pure) mathematics courses in the "20 group" have 26 to 28 percent women. In the next group of courses, which is a prerequisite for subsequent admission to a master's program in pure mathematics, the percentage remained at approximately 20 percent over the period from 1988 to 1993. This represents a small increase from the period 1982 to 1987.

What is the present situation for faculty, new PhDs and students in the PhD pipeline? Has the picture Hilmen painted in her opening address changed?

At the end of 1993, still only ten persons (7%) on the permanent staff in mathematics are women (three full professors). As in 1989, "this places mathematics in the bottom group together with physics and informatics [the Norwegian designation for computer science] as the most male-dominated research field together with technology."

In the recent years, 1989 to 1993, a total of 58 candidates obtained a PhD in mathematics. Of these six are women, making the proportion 10 percent. The corresponding percentages for physics, computer science and the field of technology are 15, 10, and 10 percent. The average percentage in 1993 for science (including mathematics) is 22 percent, which is roughly the same as the percentage for humanities and social sciences. For medicine, the corresponding figure is 38 percent and for agriculture 46 percent.

In 1989 35 percent of PhD aspirants were women. This figure has now risen by some four percentage points, as has the figure for natural sciences and for mathematics. The percentages for mathematics, physics, chemistry, biology, computer science and technology are now 20, 29, 49, 42, 18, and 19 percent respectively.

NATIONAL ACTIONS TO PROMOTE EQUITY

The national curriculum guidelines of 1974, mentioned earlier, stated that instructional materials were not allowed to give gender-restricted, biased, or discriminatory descriptions of occupations. Such materials should give realistic and varied patterns of identification for both girls and boys. But very few concrete results came out of these statements. An extensive discussion in the 1970s on women's positions in society led to the *Act on Gender*

Equality in 1979 and to the creation of a *likestillingsombud*, Ombudsman for Equity Issues, together with a Board of Appeal for Gender Equity. The law of 1979 also had a section on teaching aids, a reaction to years of biased textbooks in school, which echoed the curriculum guidelines of 1974 on the subject. The boys in textbooks were active, the girls passive. Women were invisible or tied to traditional roles. In mathematics and natural sciences, practically all examples in textbooks were gender neutral or drawn from boys' field of interests. Of course, girls have the same need as boys to be taught mathematics "in a way relevant to their experiences and lives," as Hilmen put it, and they need role models in textbooks (and in their teachers). The law does not prevent publishers or authors from issuing books which are gender discriminatory, but schools are not allowed to use such books. Every textbook used in preschool, primary, and secondary school, teacher training, and adult education is now assessed by The National Centre for Educational Resources. The authorizing system is engineered to ensure textbook quality in the areas of educational substance, language, and "equality for all categories of learners."

Every second year, an award is given to the author of the textbook that best represents a gender-equity perspective. The award was established by the Association of Norwegian Non-fiction Writers, the Ministry of Education, Research, and Church Affairs, and the Association of Norwegian Publishers. In 1987 the prize was given to a mathematics textbook in elementary school and in 1991 to a science textbook. In this context, it should also be mentioned that a competition was arranged at the beginning of the 1980s to make up the best problems to stimulate the mathematical interests of girls in the 15 to 17 age group. The suggested problems were later collected and published.

As a follow-up to the law on equity, incentive programs were initiated by the government, one in 1981 and another in 1985. In the field of education these programs focussed on teacher training, the goal being to make teachers better prepared to handle questions relating to gender equity in education. Also, informational material was distributed and special courses arranged to encourage women to apply for leadership positions in the school system.

The law introduced loose gender quotas with regard to positions in general, but allowed for strict local gender quotas with regard to recruitment of students. The Norwegian Institute of Technology seized the opportunity to introduce a bonus-point system for female applicants. Special student stipends were given by the Research Council to women in their last year of MS studies in heavily male dominated fields, including mathematics. (These stipends were subsequently divided into smaller amounts, in order to accommodate more students, and finally discontinued entirely. The reason for the latter move was reportedly that the small stipends required too much administrative work for the level of funding involved!)

In 1988 a white paper on research set as a goal that 40 percent of research stipends (mostly PhD stipends) be awarded to the under-represented gender. In particular, the Research Council took the task seriously and, as we have seen, improved figures (albeit not much in mathematics) were the result. In recent years there have been some cross-discipline university stipends for women, for example, in mathematics and computer science. Piene says in her (1993) article on female mathematicians in Scandinavia, that it is not very likely that the steps taken so far have had any influence on increasing the proportion of women in mathematics in Norway. In particular, she points out that the rule on loose gender quotas with regard to appointments has not worked particularly well in any field.

In addition to laws and rules, there has been a lot of campaigning to encourage girls to choose traditionally male-dominated studies and careers — with little effect it seems; the Norwegian labor market is still strongly gender segregated.

STUDIES ON GENDER AND MATHEMATICS

There have been no research projects on mathematics and gender initiated by the authorities in Norway. There were, however, some projects focussing on mathematics and gender around 1980, and lately there have been some signs of activity in this arena.

The results of official tests in lower secondary school and centralized examinations in lower and upper secondary school are not gender differentiated. Although it is possible to differentiate the grades in mathematics (and other subjects) on a gender basis, this is not done for some reason. Researchers have, however, obtained relevant data from lower secondary school differentiated by gender. To the best of the author's knowledge, nobody has carried out similar investigations in connection with the official examinations in upper secondary school.

Brock Utne and Haukaa (1980) obtained some sensational results in their study. Comparing the results from examinations with the average marks based on classwork, they found that girls were systematically underestimated compared to boys: the teachers expected less from girls than from boys. Imsen (1981) also conducted a large survey in lower secondary school. She did not find any significant gender differences in the marks given. Jernquist (1982) did find some differences between boys and girls in achievement in mathematics; namely, girls were not as good as boys at solving practical problems. Garmannslund (1983) commented that such practical problems had been taken from the boys' sphere of experience. Garmannslund's study focussed largely on how the different experiences and attitudes of boys and girls influenced their achievements in mathematical problem solving.

Borgersen (1984) carried out a study of a more psychological nature among pupils in Grades 7, 8, and 9. In particular, she handed out a list of 20 questions concerning feelings and attitudes towards mathematics, the usefulness of mathematics, gender roles, self-assessment, milieu, stress, and so on. The grade 7 boys and girls answered only one question differently ("Boys are better at mathematics than girls"); in Grade 8 five questions were answered differently; in Grade 9, 11 of the questions revealed significant differences between boys and girls. In particular, only four of the 64 female ninth graders in the study answered "true" in response to the statement "I think I can do more difficult mathematics." Borgersen concluded that it is in Grade 9 (the last grade of lower secondary school) that girls shy away from mathematics.

A study by Gustafsson (1993) of gifted girls and their achievement in mathematics also focussed on the last part of lower secondary education. Some of the gifted girls were underachievers for their potential, especially in mathematics, and their underachieving started in lower secondary school. Gustafsson ended up asking the question: Are gifted girls losers in the Norwegian lower secondary school?

Jernquist is at present involved in several investigations on the attitudes and achievements of pupils in lower secondary education. The thrust has so far not been on gender comparisons. In particular, she is heading a case study in connection with an Organization for Economic Co-operation and Development (OECD) project, Science, Mathematics and Technology Education in OECD Countries; the Norwegian study is concentrating on teaching and assessment in mathematics, with an emphasis on training in self-assessment. Jernquist has communicated (personal communication, 1994) some preliminary findings, which at first glance may look surprising. Two examples: 66 percent of the boys compared to 55 percent of the girls answered "yes" to the question "Do you like discussing problems and different ways of solving them in class?"; 79 percent of the boys compared to 60 percent of the girls answered "yes" to the question "Do you want a special grade telling something about effort, persistence, helpfulness and co-operative skills?" (2000 pupils in Grade 9 participated). In addition to the OECD project, Norway is also participating in the Third International Mathematics and Science Study (TIMSS).

One should also mention the book *Er skolen for Kari eller Ronny?* (Is School for Kari or Ronny?) (Gulbrandsen, 1993) that was published last year and attracted some attention in the media. The book is not a scientific study but does give a description of the regular school day as experienced by the author, who is a feminist and long-time teacher in elementary school. It contains summaries of national and international research on school and pupils, and gives practical advice to improve the situation in school for everybody. She claims that the co-educational system that was implemented in all Norwegian schools by the end of the 1950s was organized in a way that favored boys. She recommends that boys and girls be separated in some lessons so that girls can get the teacher's full attention and a greater opportunity to bring forward their experiences and opinions.

As to pupils in upper secondary school, Grønmo (1992) has studied students' choice of mathematics courses in the area of General Studies. She found that the proportion of girls in the mathematics program for the natural sciences (see Figure 2) is unchanged under New Structure policy, while the proportion of girls in the mathematics for the social sciences program increased. The social science-related courses have less status and are conceived of as easier than the natural science-related courses by all pupils.

Grønmo's work is based on an extensive inquiry among 1600 pupils shortly after they had made their choices for the second year. With the exception of those performing at the highest or the lowest level on the Norwegian seven-tier marking scale, fewer girls than boys decided to proceed with mathematics after the first year. The girls gave "math is difficult" as their main reason for quitting, while "math is not interesting" was the main reason for the boys to quit. Many girls choosing mathematics aim for a higher education within health care (MS) and medicine.

There is no finished work on assessment, mathematics, and gender from upper secondary school, but Berge (1993; Berge & Varøy Haga, 1994) has an ongoing research project on the subject. This includes material on gender differences in mathematics collected in upper secondary school. Her work shows clearly that girls' achievement in mathematics, measured by marks, drops between the first and second term of the first year. She asked pupils what they thought of their marks. The material does not uncover any significant differences between the genders as to opinions on fairness in the marks given (which is a bit surprising). But the material does indicate gender differences in looking at mathematics in connection to society. It also indicates gender differences in rural as compared to urban areas. As an extension of this work, Berge has been asking questions about different styles of assessment and their impact on the performances — and the long-term achievements — of boys versus girls (Ve, 1989). She also asks: Can the different concepts of rationality developed by Norwegian women social scientists (i.e., technical limited rationality and responsible rationality) establish a framework in which to analyse the question of gender and mathematics in a satisfying way?

In this context, it is appropriate to mention the Secretariat for Women and Research, which is an agency under the Research Council of Norway. Its objective is to increase the recruitment of women to all research disciplines and to promote, co-ordinate, and disseminate women's studies. The secretariat issues its own journal, *Nytt om kvinneforskning* (News in Women's Research). The secretariat also plays a part in arranging conferences and publishing reports. The SÅNN, JA! conference is an example of this. Meetings to discuss recruiting women to research in mathematics and physics and to computer science have also been arranged by the secretariat.

EPILOGUE

Reform 94 has already been mentioned as the latest step in the development of post-elementary education. The reform consists of more than a statutory right to education and training. It has, for instance, reduced the number of basic programs drastically. Thus, as of August 1994, General Studies and Commercial Subjects (see Figure 2) constitute one core program called General and Business Studies that attracts 60 percent of the youth. A new mathematics curriculum is now being implemented to serve the needs of this program. There will be five hours of compulsory mathematics in the first year, but to quote from the curriculum:

> In order to provide as many as possible of the incoming pupils with meaningful and appropriate instruction, the course offers, in addition to a common basic module, two alternative supplementary modules, a practical one aiming directly at the use of maths in society and daily life, and a theoretical one in which more attention is given to laying the theoretical and practical foundations for more advanced study.

The basic module is meant to give the pupils a second chance with the mathematics they were exposed to in lower secondary school and also a glimpse of some new mathematics — geometrical calculations and trigonometry. In addition, mathematics as cultural heritage, language, and communication, plus model-building and problem-solving are introduced as common targets in all modules and for all groups of pupils.

The alternative modules have caused some dispute. Some think that all pupils will take the more theoretical second module, some think *the girls* will take the easier way out. It should be emphasized that the choice is made by the pupils with a considerable amount of guidance. Moreover, the learning activities and the assessment style (more weight on projects and oral presentations) should not be to the girls' disadvantage. The Ministry has promised to follow developments by compiling appropriate statistics. A SÅNN, JA 2 conference is planned to take place in the Fall of 1995. Discussions on Reform 94 will play a prominent role.

NOTE

The author wishes to thank all those in the Ministry of Education, Research, and Church Affairs, agencies, and institutes who so readily supplied information. I also wish to thank friends and colleagues who assisted me. Above all I am grateful to Margaret Combs who patiently bore with me through several revisions of this report.

REFERENCES

Berge, L. (1993). *Kjønnsulikheter knyttet til matematikkfaget i vidaregående skule, all-menfaglig studieretning*. SÅNN'JA! rapport. SKF, Norges Forskningsråd.

Berge, L., & Varøy Haga, M. (1994). Norway. In L. Burton (Ed.), *Who counts? assessing mathematics in Europe*. Stoke-on-Trent: Trentham Books.

Borgersen, A. M. (1984). *Redd for matematikk?* Pedagogisk senter, Kristiansand.

Brock Utne, B og Haukaa, R. (1980). *Kunnskap uten makt*. Universitetsforlaget.

Doktorgradsstatistikk. Oversikt over forskerpersonale (1994). Utredningsinstituttet for forskning og høyere utdanning.

The Ministry of Education, Research and Church Affairs. (1994). *Education in Norway*. Oslo: Author.

Garmannslund, K. (1983). *Kan ikke jenter regne?* Gyldendal Norsk Forlag.

Grønmo, L. S. (1992). *Hvorfor matematikk?* Skrift nr.8. Senter for Lærerutdanning og skoletjeneste, Universitetet i Oslo.

Gulbrandsen, J. (1993). *Er skolen for Kari eller Ronny?* Ad Notam, Gyldendal.

Gustafsson, K. (1993). *Evnerike jenter og deres skoleprestasjoner i matematikk*. SÅNN, JA! rapport. SKF, Norges Forskningsråd.

Hag, K. (1994). *Tallene teller*. Norsk Matematikkråd.

Harnæs, H. og Piene, R. (1988). *Jenter i datafag og matematikk*. Nr.8 i en skriftserie fra Senter for realfagsundervisning. Universitetet i Oslo.

Holden, I. (Ed.) (1984). *Masse matnyttig matematikk*. Universitetsforlaget.

Imsen, G. (1981). *Søkelys på matematikken i ungdomsskolen*. Pedagogisk senter, Trondheim.

Jernquist, S. (1982). *Hovedoppgave i pedagogikk*. Universitetet i Oslo.

The Ministry of Education, Research and Church Affairs. (1993). *Curriculum for upper secondary education. Mathematics*. Common general subject for all areas of study.

NLS-information. (1994). The National Center for Educational Resources.

Piene, R. (1993). Kvinnelige matematikere i Norden og i Europa – i dag og i morgen. Brandell, G. et al. (Eds.). *Kvinnor och matematik*. Konferensrapport. Høgskolan i Luleå.

Ve, H. (1989). *Educational planning and rational ambiguity*. Paper presented at the VII World Congress of Comperative Education, Canada.

BODIL BRANNER, LISBETH FAJSTRUP,
AND HANNE KOCK

GENDER AND MATHEMATICS
EDUCATION IN DENMARK

Denmark is a small country both in area and number of people, with 5.2 million inhabitants. Formally there is gender equity. Nevertheless there are big differences between men and women in their choice of education, the promotion they look for or obtain, the salaries in male- and female-dominated areas, the kind of political involvement men and women choose, and so on.

There is the same elementary education for everybody, consisting of nine years of compulsory school. Most of the elementary schools are public and free of charge, and the same is true for the secondary schools, following the compulsory school. While only about 2 percent of a particular year would graduate from high school in the beginning of the century (and among those only a few women from girls' high schools) the percentage is nowadays about 45. Moreover, women constitute a slight majority of these pupils. The universities are all state universities and also free of charge for those accepted. Also at the universities there is now a slight majority of female students. But there are big differences between the various subjects: most women study humanities, fewer mathematics and natural sciences, and fewest technical subjects.

In the sixties and seventies, women in huge numbers started to become educated and to enter the labour market. Denmark is among the countries with the highest labor force participation rate for women. The total number of jobs in the country is still increasing, but there is also a large unemployment rate (the official figures for 1994: 11% in total, 10% for men, and 12% for women).

The large expansion of the universities took place in the sixties. Since then there have been fewer job openings. At the end of the next decade, we will enter a period where many more jobs have to be refilled. At the moment, there are extremely few women (17%) among the scientific staff at the universities. The changes are very slow, as has recently been documented (in Borchorst, Christensen, & Rasmussen, 1992). The universities appear to be the public institution with the least equal distribution of jobs among men and women. In two other recent publications addressing the general problems of recruitment of new research staff at the universities (Hoffmann & Poulsen, 1992; Ståhle, 1993), the special problem concerning the low number of women has also been touched on. Different conclusions were obtained: the first report recommended "a plan to make the gender distribution of the staff in better harmony

G. Hanna (ed.), Towards Gender Equity in Mathematics Education, 139–153.
© 1996 Kluwer Academic Publishers. Printed in the Netherlands.

with the gender distribution among the students of the various disciplines," while the second report concluded that "the gender distribution among the researchers and teachers at the universities are clearly undergoing changes" and it is believed the problem will resolve itself, although slowly.

Mathematics has a positive image in primary and secondary school, and many students — including many girls — like mathematics and perform well. When moving on to high school, fewer girls than boys choose the mathematical line. Nevertheless mathematics does not seem to be a major problem, it is physics which is the most gender polarizing subject. At the other end of the spectrum, the university level, we find only three women mathematicians out of a total of 121 mathematicians — that is 2 percent.

We believe that the subjects chosen by the students are greatly influenced by the educational structure, by the system of curricula options, and by tradition and culture. In the following, we will compare the gender-specific situation in mathematics with that in other subjects, beginning with the elementary school and continuing all the way to university. The figures are accompanied by brief descriptions of the school system. We shall point out where women drop out and where we see changes in this pattern. We shall also discuss some of the recent initiatives to change the situation.

THE ELEMENTARY SCHOOL, *FOLKESKOLEN*

Denmark has a nine-year compulsory co-educational school (from the age of 7 to 16 years). In principle, this means that the school is not allowed to segregate or exclude on the grounds of differences in achievement, ability, or sex. The nine years of compulsory school finish with an examination which is taken by practically everyone although it is optional. Other options include a one-year preschool-kindergarten and a tenth grade followed by an extended examination. After the ninth or the optional tenth grade, there is a choice of a two-year or three-year upper secondary school (high school), a technical college, or various types of vocational training.

Historical note

> In 1903, Parliament decreed that schools that had previously been open to boys should gradually be opened to girls and thus become mixed schools. Through the twentieth century a bizarre form of co-education emerged. Boys' schools were opened to girls, who now in one blow had to face a triple disadvantage: first a reduced curriculum in mathematics and physics (but enough in case they did not get married); second, special subjects such as home economics and needlework to qualify them as good housewives, and third, being reminded of their subordinate position to boys, whose schooling (curricula and content) aimed at the "real thing", namely to qualify them as breadwinners, and who displayed their superiority through a dominant ideology and behaviour with the support of their (mostly male) teachers. (Kruse, 1992, p. 101)

As an example of the special treatment of girls in mathematics, we find the following quote in the preface of an arithmetic book published in 1935: "Girls, having half as many lessons as boys, solve the odd-numbered problems only" (Arvin, Håstorp, Kålund-Jør-gensen, & Thorborg, 1935). The book was still officially in use as late as 1946.

Today girls and boys are generally taught together within their own age-group. But recently some innovative experiments have started teaching children in single-sex settings for a longer or shorter time, up to a whole term in certain subjects. This kind of teaching can seem very provoking only a decade after the closing of the last segregated school in Denmark, but nowadays arguments for segregation are different. "Personal development, awareness of discrimination and political understanding" are keywords in this new progressive pedagogical theory. Anne-Mette Kruse, Centre for Feminist Research and Gender Studies, University of Aarhus (1992), argues:

> Alternating between periods of mixed settings and periods of single-sex settings polarises and makes more obvious the stereotyped gendered behaviour patterns and contradictions of inequality.... Single-sex pedagogy in these projects is a means to help girls as well as boys understand sex-roles and attitudes as social constructions that can be changed by those involved.... A person is better qualified to participate on an equal footing with the opposite sex if that person has had a chance to develop according to her/his own terms. The advantage may at first appear to be greatest for girls, but in the long run both sexes can profit if boys are able to see girls contesting boys' immediate sovereignty instead of having their false assumptions reinforced. Back in the mixed class girls show less inclination to accept patriarchal values and patterns of work.... The aim is to create an awareness of our similarities and differences as human beings and an awareness of how social divisions based on race, class and sex structure our experiences and opportunities in life. (Kruse, 1992, pp. 98, 100, 101)

This technique of polarizing seems to be very successful for the 11- to 15-year olds. We also believe that older students could profit from a similar experience with shorter periods of single-sex teaching. The single-sex pedagogy has not been developed for any special topic, it has been used in the teaching of various subjects, including mathematics.

Over the last 20 years, the teaching of mathematics has changed quite a lot in *folkeskolen*. The process is explained in the report "Quality in Education and Teaching: Mathematics" (Jensen et al., 1990). A new comprehension of the "nature and role" of mathematics in elementary education was developed during the fifties and sixties. In a modern society, it is important that mathematics is "a subject for everybody — not a subject for specially selected groups of pupils or for an elite." The mathematical concepts can be dealt with on various levels and the teaching ought to be arranged in a "spiralling structure" where one returns to the same ideas again and again in more detail — and hence also in more abstract terms. As a consequence of these modifications, the attitude towards mathematics has changed rather substantially. "Mathematics has become a subject of joy." This quotation from the report made headlines in several Danish newspapers, when the report was published: "The time has passed, when fractions and prime numbers terrorised the poor pupils, today both pupils and parents are at ease and pleased with the subject.... Mathematics is now well established as a subject for all, as part of our common cultural background."

Teachers in Folkeskolen

The majority of teachers are women. Over the last ten years, the percentage of women teachers has increased from 59 percent (in 1982) to a stable 67 percent (in 1990, 1991, and 1992).

Everybody is viewed as qualified to teach mathematics in Grades 1 through 7, and approximately 40 percent of the teachers do that. Only 12 to 14 percent have special training (*liniefag*) in the teaching of mathematics. We do not know how many of these teachers are women.

The teachers are educated at special teacher training colleges. There is a compulsory course (of 100–140 hours) in mathematics for everybody. A teacher who is specially trained in mathematics has taken an additional course of approximately 350 hours. In 1993 about 18 percent of the teachers in mathematics at the teacher training colleges were women.

THE HIGH SCHOOL, *GYMNASIUM*

After elementary school, pupils can either continue in the tenth grade, start an apprenticeship, or continue in a *gymnasium* (high school) for three years. A certificate from a *gymnasium* is the normal route to higher education, especially at university. Pupils who have attended the tenth grade can continue in a two-year course known as "higher preparation," HF. Successful accomplishment, together with a well-chosen variety of subjects, may provide the same entrance qualification as a *gymnasium* education. Approximately 45 percent of young people in a particular year attend a high school or a higher preparatory course, while 45 percent start an apprenticeship.

The high school is traditionally divided into two main lines known as the "language" and the "mathematical" line.

The proportions of women students in the *gymnasium* are shown in Figure 1. The columns give the percentage in the mathematical line, the language line, and in total in the years 1982, 1987, and 1992 respectively.

Only small changes have occurred during the last ten years. At the moment, 58 percent of the students in the *gymnasium* are girls. We see that girls constitute a large majority in the language line and a smaller minority in the mathematical line. A reform in 1988 does not seem to have affected this pattern.

In the research project "Myths and realities: Women in scientific and technological education," Vedelsby (1991) has demonstrated that the choice of line in high school is determined partly by the positive interests of the students and partly by their wish to avoid physics (and technology). Mathematics does not seem to be a real problem at this stage.

Before the reform in 1988, the mathematical line of the *gymnasium* was subdivided into branches, and the courses in mathematics at the highest level were linked with courses in physics or, at some schools, with courses

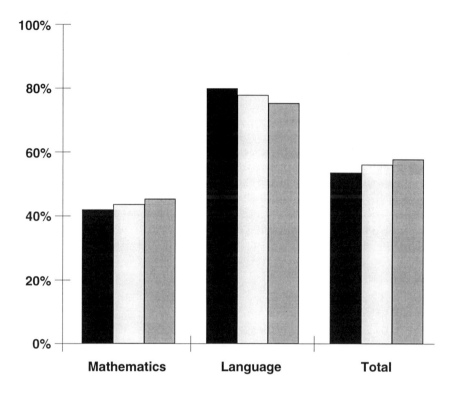

Figure 1: The percentage of women students in the *gymnasium*

(*Source: Database at the Ministry of Education.*)

in chemistry. The number of female students was always lowest in the cases where mathematics was linked with physics and higher when linked with chemistry. Since the reform, everybody in the mathematical line has two years of compulsory mathematics. In the third year, they have to choose two subjects at the so-called "high level." Almost as many as 80 percent of the math students choose the high level course in mathematics. The course includes: vectors in two- and three-dimensional space, integration, and differential equations. This course is a prerequisite for higher education involving mathematics, such as university studies in mathematics, the natural sciences, and technology. But the high participation rate also shows that mathematics is popular.

Before the reform in 1988, mathematics was also taught in the language line. Since the reform, it is no longer taught as an independent topic. Instead *naturfag* is introduced, which is a mixture of mathematics, physics, and chemistry.

144 BODIL BRANNER, LISBETH FAJSTRUP, & HANNE KOCK

Historical note
The first high schools to educate women were high schools for women only. We do not have statistical data from these early special high schools, showing their importance in encouraging women for further studies. But we have been told that at the time when these schools existed, parallel to mixed high schools, a disproportionately high number of women students in mathematics at the university had graduated from high schools for women only.

Teachers in the Gymnasium

Nearly all teachers in the *gymnasium* have a university education at the level of a master's degree, completed through half a year of pedagogical training with experienced high school teachers as mentors. In total, the proportion of women teachers is 43 percent (1994). In mathematics and the natural sciences, the percentages are lower: chemistry 30 percent, mathematics 25 percent, *naturfag* 17 percent, and finally physics with only 14 percent.

THE UNIVERSITIES

Historical note
In 1875, Danish women were given the right to become academic citizens; this means the right to graduate from a *gymnasium*, followed by the right to study at the university. Ten years later, in 1885, the first woman graduated in medicine from the University of Copenhagen.

> In 1895 Thyra Eibe graduated in mathematics from the University of Copenhagen as the first woman mathematician in Denmark.... In high school she had studied Latin and Greek. Thyra Eibe never married, but had a career as a high school teacher in private schools. Her principal achievement was the translation [from Greek] into Danish of ... Euclid's Elements.... Here she combined her interests in Greek and geometry.... The first four "books" have been reprinted as late as 1985.... She wrote several books on geometry. (Høyrup, 1993, p. 87)

> There may have been earlier women amateur mathematicians in Denmark, but at present they are unknown.

There are seven universities in Denmark: in Copenhagen, Aarhus, Odense, Roskilde, and Aalborg, moreover a technical university in Lyngby (a suburb of Copenhagen) and the Royal Veterinary and Agricultural University in Copenhagen. These all employ a scientific staff in mathematics, but a master's degree in mathematics can only be obtained at the five institutions first mentioned.

University Studies

A cand. scient. degree (a master's degree) includes a study of two subjects: a major and a minor subject. A student who chooses mathematics as one of these subjects is free to combine it with any other subject such as Danish, music, philosophy, and so on, but the usual combinations are mathematics and either computer science, physics, or chemistry. These standard combinations are promoted by the fact that one saves half a year by choosing one of them compared to a choice of an untraditional minor. During the period 1960 to 1970, there was also the possibility of taking more mathematics instead of a second subject; this attracted more women than the combination of mathematics and physics which by then was the only other combination. In Denmark, chemistry is rather popular among women, and one may expect more women graduating in mathematics and chemistry in the years to come. It was only in 1986 that this combination was made official.

A student who follows one of the standard combinations will take courses in both subjects for the first two-and-a-half years. Thereafter she or he will have to decide which subject to major in. The percentage of women students in mathematics has been rather stable over the last ten years. To illustrate the percentage of women and the differences in the choices made by women and men, we look at the numbers from the University of Copenhagen in the period 1984 to 1994. There were 32 women (31%) majoring in mathematics. Out of these, 13 had an untraditional minor subject; in comparison only 11 men chose another minor than the three traditional ones. There were eight women and 14 men with a minor in computer science, but recent numbers show that the percentage of women in computer science is dropping dramatically. The oldest combination — a minor in physics — was taken by only seven women, but by 41 men. This illustrates, again, that the lack of women is much worse in physics than in mathematics. Only two people had chemistry as a minor subject, and they were both women.

Until 1990, students could only graduate from university with a master's degree, prescribed to five years of study; now they may leave with a bachelor's degree after three years of study.

The PhD degree in mathematics (called *Licentiat* until 1991) has been awarded in Denmark only since 1972. The degree is usually achieved after two-and-a-half to three years of study on top of a master's degree. At the universities in Odense and Aarhus, this structure has been changed; since 1991 the students enter graduate school after four years of study. After two years of graduate studies, there is a qualifying exam and then after another two years they finish their degree with a thesis defence, as usual.

Positions at the University

There are three different kinds of positions for the scientific staff at the universities: *adjunkt* (assistant professor), *lektor* (associate professor), and professor (full professor). An *adjunkt* position used to be a four-year position, but is now a three-year position. A *lektor* position is permanent and so is a professor position. It is worth noticing, that the number of professorships in Denmark is very limited, so there are many *lektors* who would qualify for a professorship.

In Denmark, there is a tradition at all universities, except in Aalborg, not to count the statisticians as mathematicians; we follow this tradition and count the mathematicians as one group and the statisticians as another. Computer scientists are not included at all.

The staff in mathematics at the seven universities includes three women (one professor, one *lektor*, and one *adjunkt*) and the total number of mathematicians at the universities is 121 — that is, 2 percent are women. In statistics, there are five women (three *lektors* and two *adjunkts*) and 28 men — that is, 15 percent are women (statisticians at the Technical University are not included). With these low numbers it does not make sense to study change over the years.

In order to put these figures into perspective, we shall provide three more graphs. The first one, Figure 2, shows the percentages of women within the different faculties of the first six of the seven universities mentioned above. The faculties are: the Faculty of Arts, Theology, Social Sciences, Natural Sciences, Medicine, and Technology. For each faculty we see four columns. From left to right they indicate: women students (in the year 1980), women graduating with a master's degree (in the academic year 1986/87), women having some kind of scholarship after finishing a master's degree (in the years 1984–89), women on the scientific staff, that is, *adjunkts*, *lektors* and professors (in the year 1990).

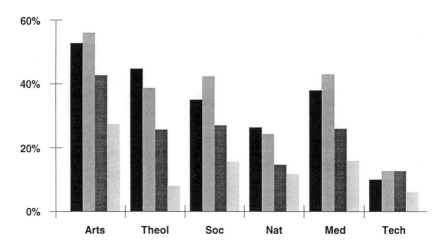

Figure 2: The percentage of women affiliated with universities

(Source: Borchost, Christensen, and Rasmussen, 1992)

Since 1980, the percentage of women students at the universities has increased to constitute a slight majority in total. To illustrate the differences within the Faculty of Natural Sciences, we show in Figure 3 the percentages of women within the various disciplines at the University of Copenhagen. The disciplines are: computer science, physics, mathematics, chemistry, geology, biology, biochemistry and geography. For each discipline, we see three columns. From left to right they indicate: women students entering in the year 1992, women graduating with a master's degree in the years 1990 to 1992, and women in the scientific staff in the year 1992.

We observe that the percentage of women is high in the "soft" natural sciences and low only in the "hard" ones. While the figures in the "hard" disciplines have been rather stable over the last ten years, the figures have increased in the soft disciplines in recent years and the number of graduates are therefore lagging behind.

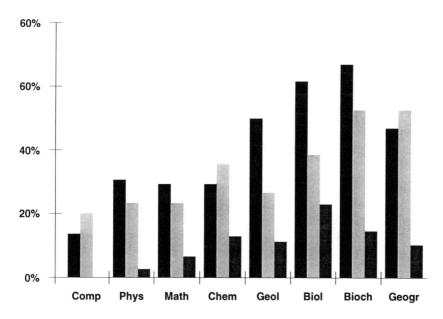

Figure 3: The percentage of women affiliated with the Faculty
of Natural Science at the University of Copenhagen

(Source: Hoffmann, Poulsen, et al., 1992)

Finally, Figure 4 shows the percentages of women within mathematics and statistics for the universities (with the exception of statisticians at the Technical University). We see four columns. From left to right they indicate: women graduating with a master's degree in the years 1984 to 1993, women who are currently PhD students, women awarded a PhD degree (in the years 1972–94), and women who are currently on the scientific staff. The figures above the columns give the actual number of women.

It has not been easy to obtain information about the total number of awarded PhDs. (For the University of Copenhagen, the number covers the period 1984–94 only.) For certain, we have not overlooked a single woman.

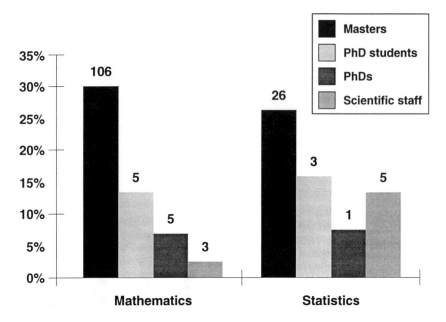

Figure 4: The percentage of women affiliated with the department of Mathematics and Statistics at universities

So with complete figures, the percentage of women would be lower than the number shown. Moreover we have not included those who have been awarded a doctoral degree abroad. Before 1972, the only way to obtain a PhD was to go abroad. Today it is no longer necessary, but still rather common.

The percentage of women obtaining a master's degree in mathematics has been rather stable over the years, and in the last ten years 30 percent of the degrees awarded have been to women. The corresponding figure for statistics is 26 percent. The proportion of women among those receiving a PhD degree in mathematics (statistics) from a university is only 7 percent (8%), corresponding to five women (one woman). Since the actual figures are so low, it is meaningless to discuss changes over the years 1972 to 1994. But the percentages of women who are currently PhD students are in fact significantly higher: 14 percent in mathematics and 16 percent in statistics, representing five women in mathematics and four in statistics.

Nevertheless we observe that the main drop in the percentage of women occurs between the level of a master's degree and entering a doctoral program. One may wonder whether the main reason why women are involved in a PhD program in larger numbers now is just because the age of having children has changed. If this is true, an essential rise in the number of women faculty will not occur until we start having our children when we are close to 40.

Some initiatives have been taken to increase the number of PhD students.

Since April 1994, there has been a change in the length of maternity leave for women students in general. The normal length of maternity leave for any professional woman is between 28 and 32 weeks, with a salary corresponding at least to maintenance allowance. Most women in academic positions keep their full salary, though. Ten of these weeks may be transferred to paternity leave. For students, the rules have improved substantially. A woman student can get a full year of maternity leave. A PhD student keeps her full scholarship during this period. A male student may add another half year of paternity leave. Or vice versa: a couple of students may choose half a year of maternity leave, followed by one year of paternity leave.

No initiatives have yet been taken to increase the number of PhD students, post-docs., and so on in mathematics in particular. The recent initiatives are broad and directed towards both mathematics and the natural sciences. The Danish Research Council of Natural Science has set aside a PhD stipend for women only. (That means one stipend out of 15 every half year.) The number of female applicants has risen dramatically, and the women candidates have proved to be as qualified as the men and have therefore — as a group — obtained more that this one stipend. However, the special stipend has had an encouraging effect. In 1994, three post-docs. for women only were offered by the Faculty of Natural Sciences at the University of Copenhagen. The initiative taken by the university was met with resistance from the Ministry of Education since it violates the Act Against Gender Discrimination, but in the end the Ministry was overruled by the support from the National Council of Equity. More offers of this kind are expected to occur in the near future.

In 1994, all the universities were asked by the Ministry of Education to formulate plans of action for creating better equity. It remains to be seen if anything will come out of this. We do believe that some special initiatives are necessary.

THE TRADITION OF FOLK HIGH SCHOOLS

Besides the types of schools already mentioned, there is a further group which is of considerable national importance, known as the Folk High Schools. These are non-vocational schools; they do not prepare for examinations but impart general adult education. The teaching is based on the ideas suggested by the great 19th-century theologian, poet, and educator N.F.S. Grundtvig, who said that a condition for democracy was that each citizen should be informed of the country's history, literature, and social conditions.

Historical note
According to Grundtvig, mathematics was not part of "education of the people," but together with Latin "soul-destroying and fatal."

Mathematicians preach a pure science, which in principle neither pays attention to life nor death or any other kind of human activity, and even when applied is so universal and cosmopolitan that it cannot possibly confine itself to any language, or serve the interest and common good of any particular people or country. (Hansen, 1983, pp. 97, 98)

In the 1870s, more advanced courses were introduced at the Folk High Schools, and mathematics was added to the schedule, mainly as a historical discipline. Equations, abstractions, and analytical thinking had their place, but had to be constantly guided by "the living word." A point of view that nowadays has gained new importance.

THE MATHEMATICS INITIATIVE

In 1988, the Danish Research Council of Humanities (*Statens humanistiske forskningsråd*) launched a five-year project with the title "Mathematics Education and Democracy in Highly Technological Societies." The overall intention was to discuss mathematics education as part of a democratic endeavour in a highly technological society. The aim was to support research *about* mathematics: What is mathematics? Who owns it? What can it be used for? How is it used and abused? What are the problems in teaching mathematics? How do we improve public awareness of mathematics? And how do we stimulate the public to take an interest? Comprehension came to be the key word for numerous discussions between mathematicians and non-mathematicians and among teachers from different levels of mathematics education. Results are summarized in the publication *A Gap in Culture* (Nissen, 1994).

Where does gender come in? It hardly does, except as part of a hidden agenda. Looking back at the list of publications of the project it becomes clear that "even the most perceptive and humanistic works on the relationship between teaching, culture and society rarely mention gender" (Wajcman, 1993, p. 15).

Mathematics is still considered a gender-neutral subject.

GENDER STUDIES AND MATHEMATICS

Inspired by the Swedish conference *Kvinnor och Matematik* (Women and Mathematics) in Malmø in April, 1990 a similar conference *Kvinder og Matematik* (Women and Mathematics) was held in Copenhagen during a weekend in September, 1991. Eighty mathematicians and teachers of mathematics from all levels of education discussed gender and mathematics, not as a problem, but as a matter of fact. The conference on the whole showed a tendency to work with pupils as individuals. Gender was (and is) rarely used as an analytical tool in accounts of mathematics. Although many useful contacts were formed, there are no immediate plans for a second conference.

Within the humanities, social sciences, and medicine, however, gender studies are relatively more common. Centres for feminist research and/or milieu for feminist activities have been established by most universities.

In 1990 *Foreningen for Kvinde- og Kønsforskning i Danmark* (the Association for Women's and Gender Studies in Denmark) was founded. The association aims at initiating networks within fields of women's studies, supporting the recruitment of a new generation of feminist scholars, lobbying both public and private research institutions, doing fund-raising, and acting as a scientific forum for all feminist scholars in Denmark. Together with a national part-time co-ordinator of women's studies, the association publishes a national journal for feminist research, *Kvinder, Køn og Forskning* (Women, Gender, and Research), a journal that also is a newsletter and a forum for scientific discussion. The first issue of the 1993 volume was devoted to the subject "numbers" with interdisciplinary papers such as "The Power of Numbers," "Logic and Postmodernism," "Counting Works in Poetical Praxis," and "Behind the Numbers and into Mathematics."

REFLECTIONS

There has been very little focus on "gender and mathematics" in society; it is not considered to be a problem in the school system up to university level. The gender polarization becomes first noticeable at the entrance to university, where women seem to avoid the hard disciplines: mathematics, physics, technology, and, in particular, computer science. Later at the entrance to a PhD program, the polarization is more distinct and even more so when passing to a research career at universities. At all levels there is a need for more encouragement of women and more awareness among the teachers and the authorities.

Teaching for shorter periods in a single-sex setting, alternating with a mixed setting, has proved very successful in *folkeskolen*. After some years of experimental teaching, this is now becoming more common. There is little teaching, however, in a single-sex setting at high school level; exceptions include some physics laboratory classes. We believe that it could be most valuable to extend the teaching in single-sex settings to the high school and in particular to the hard disciplines.

The choice of subjects both in high school and at university is greatly influenced by the educational structure and the system of curricula options. The mere existence of soft and hard parallel options creates a gender polarization.

The percentage of women on the scientific staff at universities is changing very slowly. In all faculties, except those in natural sciences, there has been a small increase. The post-doc. positions for women, recently offered by the Faculty of Natural Sciences at the University of Copenhagen, are an attempt to change the tendency of even fewer women than before applying for vacant assistant professorships. These post-doc. positions are also meant as an opportunity for women to re-enter a university career if they have left after a PhD degree. We believe that the situation will not be resolved by itself, but that special initiatives of this kind are needed.

Each society or subculture invokes methods to ensure the reproduction of gender roles. In modern Western societies with a relatively high degree of equity in public life, with both wives and husbands active as breadwinners, the mechanisms repressing either gender can be rather disguised. Each historical-cultural period must evaluate the prevailing gender roles (and related myths) and judge whether the restrictions put on men and women are acceptable or not. (Beyer, in press)

NOTE

We wish to thank all those who supplied information: the Ministry of Education, the Administration Offices of the different universities and institutes, LMFK (the Association of High School Teachers in Mathematics, Physics and Chemistry), SEMAT (the Association of Teachers in Mathematics at Teachers Training Colleges). We are grateful to Mette Vedelsby for her support and valuable comments.

REFERENCES

Arvin, G. J., Håstorp, H., Kålund-Jørgensen, Å., Thorborg, S. F. (1935). *Regnebog for folkeskolen i København , 8. hæfte* [Arithmetic book, part 8]. København: Ascherhoug Dansk Forlag.

Beyer, K. (in press). A gender perspective on mathematics and physics education: Similarities and differences. In B. Grevholm (Ed.), *ICMI study 1993/ Gender and mathematics education.*

Borchorst, A., Christensen, A.-D., & Rasmussen, P. (1992). *Kønsprofilen i videnskabelige stillinger ved de videregående uddannelser* [The gender profile in scientific jobs of higher education]. København: Statens information.

Hansen, H. C. (1983). Poul la Cour – en grundtvigsk naturvidenskabsmand. In *Fremtidens Videnskab. En debatbog om Grundtvigs videnskabssyn og vor tid* [Science of the future. A book discussing the view upon science of Grundtvig and our time]. København: Samlerens forlag.

Hoffmann, E., & Poulsen, O., (1992). *Forskerrekruttering I og II* [Recruitment of researchers I and II]. København: Det Kongelige Danske Videnskabernes Selskab [The Royal Danish Academy of Sciences and Letters].

Høyrup, E. (1993). *Thyra Eibe - Danmarks første kvindelige matematiker* [Thyra Eibe — The first woman mathematician in Denmark]. *Normat. Nordisk matematisk tidsskrift, 2,* 41–44.

Jensen, H. N., (1990). *Kvalitet i uddannelse og undervisning: Matematik.* [Quality in education and teaching: Mathematics]. Undervisnings- og forskningsministeriet [The Ministry of Education]. København: Statens informationstjeneste.

Kruse, A. M. (1992). "... We have learnt not just to sit back, twiddle our thumbs and let them take over." Single-sex settings and the development of a pedagogy for girls and an pedagogy for boys in Danish schools. *Gender and education, 4*(1/2), 81–103.

Nissen, G. (1994). *Hul i kulturen* [A gap in Culture]. København: Gyldendal.

Ståhle, B. (1993). *Forskningspersonale og Forskerrekruttering på de højere uddan-

nelsesinstitutioner i Danmark [Research staff and recruitment of researchers at institutions for higher education in Denmark]. Forskningspolitik 13. Forskningspolitisk Råd, Undervisningsminiseriet.

Vedelsby, M. (1991). *Myter og Realiteter: Kvinder i naturvidenskabelige og teknologiske uddannelser* [Myths and realities: Women in scientific and technological education]. København: Forskningspolitisk Råd.

Wajcman, J. (1993). *Feminism confronts technology.* Cambridge: Polity Press.

LENA M. FINNE

GENDER AND MATHEMATICS
EDUCATION IN FINLAND

The purpose of this paper is to provide an overview of mathematics education in Finland, with a focus on gender. In order to do this, it is necessary to provide first the historical background and development of the Finnish education system.

According to Husu (1991), a "double strategy" has characterized the relation of Finnish women's studies to mainstream research. As separatist standpoints have never been very popular in Finland, researchers and students have tried to influence and change their own disciplines from within (Husu, 1991). This double strategy can also be discerned in the writing that has been done on gender and mathematics education. As can be seen in the list of references, there are few specific studies and articles by Finnish authors on this topic. At the same time, gender is frequently discussed in reports on mathematics and mathematics education — for example, the final report of the Committee for General Knowledge in Mathematics and Natural Sciences (the Leikola Committee) stresses the importance of giving all pupils equal opportunity to learn mathematics and natural science (Kommittébetänkande, 1989).

As there has been little research done explicitly on mathematics education and gender in Finland, the content of this article has been compiled from many different types of sources. All the information I would have liked to include has not been available. I have refrained from making comparisons to other countries. All references are either by Finnish authors or published in Finland. Some of the references are available in English.

FINLAND IN BRIEF

The Republic of Finland has a population of 5.0 million and an area of 338,000 square kilometres. This make an average density of 16 inhabitants per square kilometre (Ministry of Education, 1992). The official languages are Finnish and Swedish. Swedish is spoken by 5.9 percent of the population. Lappish is spoken by 0.03 percent of the population (Statistics Finland, 1993a). Laps are allowed by law to use their language in court and official institutions (Onikki & Ranta, 1992).

Approximately 700,000 people are of age 65 or more. The prognosis is that by year 2030 the number will have grown to 1.1 million. In 1990, the

G. Hanna (ed.), Towards Gender Equity in Mathematics Education, 155–178.
© 1996 Kluwer Academic Publishers. Printed in the Netherlands.

number of children age 7 to 15 was 588,468 and by year 2020 this number will be 484,045. In 1992, about 47,000 foreigners lived in Finland (Statistics Finland, 1993d).

The state churches are the Evangelical Lutheran and the Finnish Orthodox. About 87 percent of the inhabitants belong to the Evangelical Lutheran and 1.1 percent to the Orthodox Church (Statistics Finland, 1993a).

In 1906, all Finnish citizens were granted suffrage, which meant that Finnish women were the first in Europe to be accorded these rights. Finnish women were also the first in the world to be eligible for elected office (Manninen, 1993). At the time of the reform, Finland was an autonomous Grand Duchy of the Russian Empire; just a decade later, in 1917, after one century of Russian domination and six centuries of Swedish rule, Finland gained independence.

Unmarried women in Finland, aged 25 or over, had been permitted to manage their own affairs since 1864. In 1929, married women achieved independence under the law. Over 100 years later, an Equality Act (1987) was passed to prevent discrimination (Manninen, 1993); this law is now being revised as it has not brought the desired equality between women and men ("Tasa-arvolakiin," 1994).

Legislation gives women the right to a maternity leave up to 105 days plus a parental leave of 170 days (Rantalaiho, 1993). During these periods, they are paid part of their salary and maternity benefits in the form of a daily allowance. There is also paternity leave intended for fathers and it is possible for the father to stay at home instead of the mother. In this case, the maternity benefit is paid to the father (Manninen, 1993). Children under three must be provided with municipal daycare (Rantalaiho, 1993).

Women form half of the labor force and almost all of them work full time; 10 percent of women work part time compared to 4 percent of men. The Finnish labor market is segregated by gender (Aitta, 1988; Anttalainen, 1986; Martikainen & Yli-Pietilä, 1992; Rantalaiho, 1993). More than 90 percent of the population work in a profession where more than 60 percent are of the same gender (Nummenmaa, 1989, 1990). Women on average earn 25 percent less money than men. The smallest differences (13%) are between those who have completed research education (Statistics Finland, 1993g).

The unemployment rate was 19.5 percent in May 1994. Of the unemployed, 57 percent were men and 43 percent were women ("Työttömien määrä," 1994). For the moment, Finland is living through an economic crisis.

HISTORICAL OVERVIEW OF THE EDUCATION SYSTEM

During six centuries, until 1809, the main part of today's Finland formed the eastern part of the kingdom of Sweden. As early as the 14th century, Finnish students went to Paris, France, and Germany to study (Raivola,

1989). In the 13th century the Cathedral School of Turku was founded. This school was transformed into a university in 1640, the Royal Academy of Turku. By the 18th century the natural sciences — physics, chemistry, biology, and mineralogy — achieved pre-eminence in the University (Klinge, 1970).

As a consequence of the Reformation, being able to read and learn the catechism were considered prerequisites for Holy Communion and legal marriage for both women and men. In the yearly catechetical meetings in the villages, women received high marks and on this basis were appointed to teach not only their own children to read but others as well (Sysiharju, 1984).

In the Nordic War, the Wiborg province in the eastern part of Finland was transferred to the rule of the Russian Empire. In 1788, the Empress Catherine the Great initiated, among other educational reforms, the founding of the *Demoisellen-Classe* which rapidly grew into a separate and highly appreciated *Töchter-Schule*. Before 1811, seven other *Töchter-Schule* were founded for the bourgeoisie in the towns of the Wiborg province (Sysiharju, 1984). Among the subjects taught was arithmetic (Hakaste, 1988).

In 1809, after the war between Sweden and Russia, the whole of Finland was transferred from Swedish rule to the Russian Empire and Finland was declared an autonomous Grand Duchy. In 1811, Wiborg was unified with the new Grand Duchy of Finland. This meant that two traditions were combined in the Finnish school system.

In 1843, the 1724 statutes were replaced (Hakaste, 1988; Sysiharju, 1984). These statutes were the first in Scandinavia to recognize the principle of and the need for state girls' schools. The schools were, by law, mainly intended for girls of the middle and upper classes; the natural sciences were not part of the curriculum (Hakaste, 1988). The private schools founded by Sara Wacklin and Odert Henrik Gripenberg should be mentioned as predecessors of education for girls with a broader curriculum. At least, in Sara Wacklin's private schools, arithmetic was part of the curriculum (Mäkelä, 1991).

In 1863, the first state teacher-training college for elementary schools was founded in Jyväskylä in 1863. Right from the beginning it was a dual college (Sysiharju, 1984). Also in 1863, a declaration was made that the Finnish language should have the same official status in Finland's administration as the Swedish language; this was to take place over the next 20 years (Sysiharju, 1984). The language of instruction in secondary schools had been Swedish, and before 1850 attempts to found secondary schools in the Finnish language had failed. From the end of the 1850s to the beginning of the 1870s, there were some bilingual secondary schools. By the 1870s, the schools were divided into Finnish-language and Swedish-language schools (Hakaste, 1992).

The statutes concerning rural elementary schools came into effect in

1866 (Sysiharju, 1984). The first co-educational secondary schools were founded in the 1880s (Hakaste, 1988). There was a parallel school system, folkskola and samskola, until it was decided in 1968 to go over to the nine-year comprehensive school (Antikainen, 1983). By the 1970s, about 80 percent of schools were co-educational (Sysiharju, 1984).

In 1870, the first woman took the matriculation examination (Haataja, 1989); in 1878, the first female physician took her degree at the university; and in 1883, the first woman took a master's degree (Sysiharju, 1984; Wilkama, 1938). Although the university had been willing to accept female students on an equal basis (Sysiharju, 1984; Wilkama, 1938), the Russian Emperor had, until 1901, required individuals to apply for an "exemption based on sex." In 1891, there were 44 female students at the University; in 1901 there were 385 (16%). By 1911, about one-fourth of the students were women (Klinge, 1970; Sysiharju, 1984). A total of nine PhD's were taken by women at the university from 1895 to 1917 (Kivikkokangas-Sandgren, 1989).

In 1917, Finland succeeded in declaring itself a sovereign state. The next year, in 1918, Åbo Akademi was founded; previously there had been only one university in Finland, the University of Helsinki, which had been founded in Turku in the 15th century and moved to Helsinki by the Emperor in 1827 after the conflagration in Turku. In 1922, Turun Yliopisto was founded.

In 1992, there were 1,734 comprehensive schools, 467 senior secondary schools, 570 vocational and professional education institutions, 20 universities and university-level institutions, and 24 other comprehensive schools and senior secondary schools. In these schools, there were altogether 1,041,460 students; 51.6 percent were female (Statistics Finland, 1993d). All these schools are co-educational.

Compulsory primary education was laid down by law in 1921. Today the literacy rate is almost 100 percent. Compulsory education applies to all between the ages of 7 and 17 by present law. This requirement can be completed by attending a nine-year comprehensive school (Manninen, 1993).

THE DEVELOPMENT OF FINNISH EDUCATIONAL POLICY

Finnish educational policy after World War II has sought to raise the overall level of education. The goal has been to create equal opportunities for all, regardless of their place of residence, financial standing, language, or sex (Ministry of Education, 1992).

In 1950, 0.3 million people had taken a post-compulsory degree; in 1991, 2.0 million people had taken a post-compulsory degree (Statistics Finland, 1993d, 1994). In 1991, 52 percent of all men and 50 percent of all women had taken a post-compulsory degree. This means that the educational level has increased by 21 and 23 percentage points, respectively, for men and women since 1973 (see Figure 1) (Statistics Finland, 1993a; Tilastokeskus, 1975) and that women under 45 years of age are now more educated than men (Statistics Finland, 1993b).

Among the steps to make education more equitable, we should include the

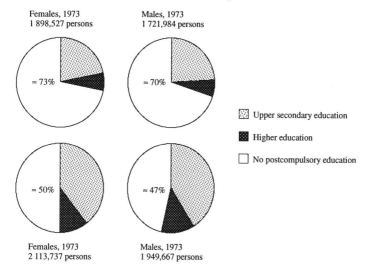

Figure 1: Population (persons aged over 15 years) by level
of education and sex in 1973 and 1991

(Sources: Tilastokeskus, 1975, p. 317; Statistics Finland, 1993a, p. 424)

introduction of the comprehensive school system, increased social support for students, and the regional decentralization of higher education to facilitate access (Lähteenmaa & Siurala, 1992). The objective of the current educational development plan, approved by the government for 1991 to 1996, is to raise educational standards, revise educational contents, and create more options and individualized programs (Ministry of Education, 1992).

Pre-primary education is provided in children's daycare centres (kindergartens) and in comprehensive schools. There are no pre-primary schools per se in the Finnish education system (Statistics Finland, 1991). In 1987/88, 34 percent of all children of age four took part in pre-primary education (Ojala, 1993).

The regular education system is composed of comprehensive schools, senior secondary schools, vocational and professional education institutions, and universities (see Figure 2) (Statistics Finland, 1991). Compulsory schooling begins in the autumn of the year when the child turns seven and lasts for ten years, unless the pupil completes the set curriculum earlier (Ministry of Education, 1992). There has been some discussion to start formal education one year earlier (e.g., Carlson, 1974; Ministry of Education, 1992). The comprehensive school provides all school-leavers with the same eligibility for further studies and is free of charge to all pupils (Ministry of Education, 1992).

In the 1980s, the drop-out rate in senior secondary schools was less than 7 percent; in vocational schools, around 8 percent. In comprehensive schools, the drop-out rate is close to zero (2 per 10,000) (Lähteenmaa & Siurala, 1992). At the universities, the drop-out rate has varied between 5.8 and 6.7 per 100 students in the period 1985 to 1990 (Statistics Finland, 1993d).

In addition to the regular education system, formal education is also given in music schools and colleges and institutions of physical education. Adult education is provided by "folk" high schools, adult education centres, and summer universities. The military academies and military vocational institutes are not classified as belonging to the education system (see Figure 2) (Statistics Finland, 1991).

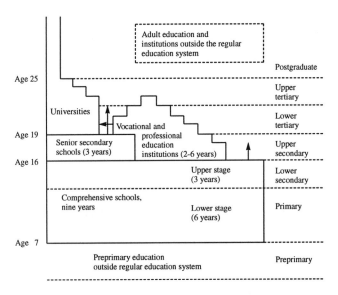

Figure 2: The Finnish Education System

(*Sources: Ministry of Education, 1992, p.7; Statistics Finland, 1993d, p. 212*)

TEACHERS

At the lower levels of the comprehensive school (Grades 1-6), pupils are taught by classroom teachers; at the upper levels (Grades 7-9), they are taught by subject-specialist teachers (Ministry of Education, 1992). Teachers in comprehensive and upper secondary schools and some full-time teachers in vocational training and adult education centres receive their pedagogical training in universities. In higher education and in some fields of vocational training, no formal pedagogical training is required (Ministry of Education, 1992). Starting in August 1995, pre-primary school teachers will be educated at the universities (Backman, 1994).

In 1991, 65.9 percent of teachers in comprehensive schools were women. In the same year, women comprised 60.0 percent of the teachers in senior secondary schools, around 52.0 percent in the vocational and professional schools, and 33.6 percent at the universities (Statistics Finland, 1993d). The female proportion of the teaching staff has grown from 61.3 (1981) to 65.9 percent (1991) in comprehensive schools and has been stable around 60 percent during the same period in senior secondary schools (Statistics Finland, 1993d). Of those taking their master's degree combined with a teacher's degree in mathematics, physics, and chemistry in 1989, 59.2 percent were female (Statistics Finland, 1991). Subject specialists in Finnish comprehensive and senior secondary schools teach all or a combination of mathematics, physics, chemistry, and computer science. According to a study done in 1990, with a sample of 49 mathematics teachers, 41 percent are female (Saranen, 1993). The corresponding figures in 1980/81 and 1983/84 were 33.5 and 33.7 percent (Sysiharju, 1984).

There has been an increase in academic career opportunities for women (see Table 1).

Table 1: Some data on women working at universities, in percent

Year	Full professor %	Professor %	Lecturer %	Assistant (two types) %	Teacher (full time) %
1973	3.8	6.6	24.2	14.1 & 22.6	–
1980	5.7	8.2	31.9	18.7 & 28.2	–
1988	9.6	12.5	44.1	22.8 & 34.4	57.3
1992	11.2	19.2	46.3	29.1 & 36.4	62.0
1993	11.6	22.1	52.2	27.6 & 27.6	57.9

Note: The group of full time teachers at the universities is comparatively new.

(Sources: Bergman, 1985, p. 24; Tasa-arvo toimikunta, 1990, p.23; Opetusministeriö, 1993, p. 45)

In 1980, 4.7 percent of the mathematics teaching force at the universities were female (Kommittébetänkande, 1982). In 1990, the total number of professors in mathematics and statistics was 46 (Pylvänäinen, 1992). In 1994, there was at least one female professor in mathematics in Finland — in applied mathematics at the University of Turku. There are, however, women working as lecturers, senior assistants, and assistants at departments of mathematics and in mathematics education at departments of education (Suomen valtiokalenteri, 1992). In comprehensive and senior secondary schools, girls have female role models in mathematics and natural sciences.

POSTCOMPULSORY EDUCATION IN THE
REGULAR EDUCATION SYSTEM

The senior secondary school provides three years of general education and ends in a national matriculation examination which gives general eligibility for higher education (Ministry of Education, 1992). The matriculation examination is the first national assessment. Before that, the evaluation of pupils' achievement is made individually by teachers (Lahelma, 1993). In the matriculation examination, there are four compulsory subjects: mother tongue, the other official language, a foreign language, and either mathematics or realia. Realia consists of history, scripture, biology, geography, physics, chemistry, psychology, and philosophy; all these subjects are included in the same test and students can choose freely which questions to answer. In senior secondary school, students can choose between an extended and a short course in mathematics. If students take an extended mathematics course it is compulsory in the matriculation examination (Statistics Finland, 1991). The senior secondary school and the matriculation examination are undergoing some changes that will give students more freedom to choose which courses they want to take and when, and what they want to write on in the examination.

In autumn 1992, about half of those beginning their studies in senior secondary school studied extended mathematics. Of these, 70 percent were boys and 30 percent were girls (Statistics Finland, 1993c). In spring 1988, 55 percent of all boys and only 22 percent of all girls taking part in the matriculation examination took the test for a degree in the extended course of mathematics (Lähteenmaa & Siurala, 1992).

Vocational, technical, and professional education are mainly provided in specialized institutions (Statistics Finland, 1991). College and higher college diplomas provide the same eligibility for higher education as the matriculation examination (Ministry of Education, 1992). In 1992, there were 201,053 students at these institutions, of which 54.6 percent were women (Statistics Finland, 1993f). Women took more than 90 percent of all degrees in the fields of nursing and social services in 1991. One can clearly distinguish between male and female fields of education, and the changes compared to 1985 are minimal (Statistics Finland, 1993d).

The university system comprises 17 scientific institutions of higher education, of which 10 are multi-faculty universities, three technical universities, three schools of economics and business administration, and one a veterinary college. There are also three art academies. Mathematics is taught at the universities of Helsinki, Joensuu, Jyväskylä, Kuopio, Oulu, Tampere, Turku, and Åbo Akademi. Mathematics is also taught at five universities (or faculties) of technology and three schools of economics and business. All the universities provide undergraduate and postgraduate education, confer doctorates, and are required to carry out research. It takes six to eight years to complete the first degree ("candidate" = master's degree), while postgraduate degrees (licentiate, doctorate) take several more years (Statistics Finland, 1991).

In 1992, the median age for a postgraduate degree was 34.9 years, 34.4 for men and 36.7 for women (Statistics Finland, 1993e). Of those accepted to the universities in 1992, 77 percent had taken the matriculation examination, 22 percent had a vocational degree, and 1 percent had no degree (Kaipainen, 1994). In autumn 1993, there were 121,200 students at the universities; 52 percent were females. Of the new students, 17,900, 56 percent were female.

In the field of natural sciences, there were 16,537 students; 43.8 percent were females. In 1992, in all fields of education, except engineering (where 18% are female), at least 40 percent were female (see Figure 3) (Statistics Finland, 1993e).

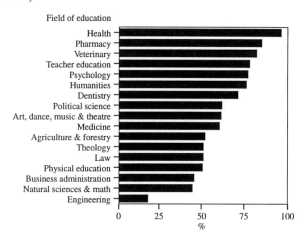

Figure 3: The proportion (in percent) of women by field of education at the universities in 1992

(Source: Statistics Finland, 1993e, p. 8)

In 1992, 41 percent of all research students were women. Of those who took licentiate degrees in 1992, 33.6 percent were female and of those who took doctoral degrees, 30.5 percent were female. The corresponding percentages in 1985 were 30.5 and 27.5 (Statistics Finland, 1993e). In 1991, 33.8 percent of licentiate degrees and 33 percent of doctoral degrees in the natural sciences were taken by women. Of those who have completed education in research, 90 percent of women and 95 percent of men belong to the labor force (Statistics Finland, 1993d).

All universities are state owned and education is provided free of charge, but each student must pay a nominal union fee. Subject to certain restrictions, university students may receive financial aid from the state (Statistics Finland, 1991).

MATHEMATICS EDUCATION

In 1988, the Leikola Committee publicly reported on the situation in mathematics and natural sciences, and in 1989 they presented their final report (Kommittébetänkande, 1988, 1989). In 1994, a new curriculum for comprehensive schools was published by the National Board of Education (Utbildningsstyrelsen, 1994). This curriculum has been criticized for not being radical enough (e.g. Haapasalo, 1993); the "old" mathematics curriculum was criticized for not promoting equality for girls and boys (Lahelma, 1993). There is also a new curriculum for senior secondary schools.

There has been research on the reproduction cycle and latent curriculum in the school setting (e.g., Meri, 1992; Metso 1992) and in the university setting (e.g., Antikainen, 1983; Liljander, 1991). Researchers have stressed the negative impact of gender-stereotyped textbooks (e.g., Lahelma, 1989), and have pointed out that research needs to be done which looks at gender socialization in different social classes (Lahelma, 1987).

COMPREHENSIVE SCHOOLS
AND SENIOR SECONDARY SCHOOLS

The National Organization for Teachers in Mathematics and Science organizes Grade 9 mathematics competitions every year for both comprehensive and senior secondary schools. During the last five years, girls have represented about a quarter of the 20 best students in the competition; there have been 1 to 4 girls among the 20 to 27 best in competition's basic series and 0 to 4 girls among the 20 to 30 best in the open series (Dimensio, 1989, 1994a, 1994b; Latva, 1991; Seppänen, Toivari, Pippola, & Norlamo, 1992; Toivari, 1990). In 1992 and 1993, there has been one girl on the Finnish Mathematics Olympics Team (Lehtinen & Saksman, 1992; Lehtinen, 1993).

Until the middle of the 1980s, it was possible to choose between different courses in mathematics in Grades 7 to 9. In autumn 1977, the extended intermediate course was chosen by 88.5 percent of girls and 80.0 percent of boys; in autumn 1980, the corresponding numbers were 85.8 and 76.4 percent (Statistics Finland, 1980, 1984). According to a study conducted in 1979, there were no differences between boys and girls in Grade 4 and small differences in favor of girls in Grade 6 and in Grade 9 in comprehensive schools; however, in the extended course in mathematics in the last grade of senior secondary school, the differences favored boys (Kupari, 1983a, 1983b, 1986). In a follow-up study done in 1990/1991, both boys and girls had achieved better results in mathematics (Kupari, 1993).

According to the national results of the Second International Mathematics Study (IEA/SIMS), the performance level by girls in Grade 7 was slightly higher than boys. But in the third grade of senior secondary school in the extended course in mathematics, boys did better than girls in various subdomains of mathematics and in tasks requiring different cognitive levels of mathematics thinking. Only in the subdomain of algebra, was the difference between boys and girls not statistically significant (Kangasniemi, 1989). When the results of a test in the ninth grade in five Swedish-language comprehensive schools were analysed, the difference between boys and girls was not statistically significant. But the girls preferred familiar tasks and the boys showed a greater willingness to try unfamiliar tasks (Björkqvist, 1994). In the same test in Finnish language schools, the difference between boys (483 boys) and girls (486 girls) was statistically significant and favored boys. Girls preferred familiar tasks and in these they were as successful as boys. But boys chose easy tasks that had not necessarily been taught at school (Korhonen, 1994). So both in the Swedish and the Finnish language schools the pattern of different task-choosing behavior was recognized.

Linnanmäki found in her sixth grade study significant differences in two tasks. Girls did better than boys in a multiplication task and boys did better than girls in a task testing number sense (1990). When testing an environment for constructivist learning of fractions, another researcher found that the girls outperformed boys but that the differences between boys and girls were smaller in the test group than in the comparison group (Haapasalo, 1992).

Kupari's study also showed that the liking for mathematics declines the older the pupils get, but that boys and girls like mathematics in all classes about the same amount (Kupari, 1986). In the follow-up study, they still liked mathematics more in the fourth grade than in the ninth grade (Kupari, 1993). Karjalainen found a somewhat contradictory result when studying pupils in Grades 1 to 6 in comprehensive schools. Girls in the fifth and sixth grades were more motivated to study mathematics than boys (1982). The results from IEA/SIMS showed that girls made a greater investment in mathematics than boys, especially in senior secondary school, and that girls had a weaker math self-concept than boys, even in the seventh grade in comprehensive school where girls outperformed boys. Girls got more help in mathematics and were encouraged by their parents to do mathematics. Boys' parents expected boys to do well in mathematics (Kangasniemi, 1989).

Pehkonen has investigated the conceptions of seventh-graders concerning mathematics and mathematics teaching. He found that girls stressed affective components while boys stressed understanding; boys disliked taking notes more than girls and would have liked to have more computers and learning games than the girls did (1994). Girls were more ready to work seriously with mathematics than boys and they disagreed more strongly with the statement that only mathematically talented pupils can solve problems. Boys were more in favor of geometry than girls (Pehkonen, 1992).

When comparing results on familiar mathematical tasks, researchers will not find any big differences between boys and girls in comprehensive schools. But they will find different patterns of task-choosing behavior, with boys being more willing to choose unfamiliar tasks and more likely to be successful in doing these tasks. There will also be gender differences in preferred working and learning methods and in attitudes to learning mathematics. And there will be a difference in math self-concept. On the surface, it looks good, but why the apparent gender differences in senior secondary schools?

CHOICES MADE AFTER COMPREHENSIVE SCHOOL

The first choice concerning mathematics is made after comprehensive school. Let us take a closer look at the year 1991 (see Figure 4).

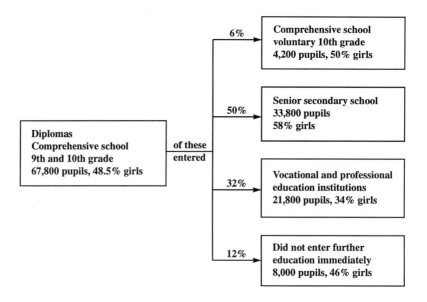

Figure 4: Choices made by the students that
completed comprehensive school in 1991

(Source: Statistics Finland, 1993d, pp. 84-85)

Most of the students completing comprehensive school in 1991 entered either vocational and professional education institutions (32%) or senior secondary schools (50%).

The fields of education in vocational and professional education institutions are segregated according to gender (see Table 2).

Of those who completed senior secondary school in spring 1991, 22 percent entered vocational and professional institutions the next autumn while 19 percent entered universities (see Figure 5).

Table 2: Leaving certificates issued and final examination taken
at vocational and professional education institution
by field and gender in 1991

Field of education	total	females %
Male dominated professsions (males 61–90%)		
Transport and communications	800	12.5
Engineering	17,235	16.5
Agriculture and forestry	3,117	33.2
Female dominated profession (females 61–90%)		
Commerce and business administration	12,946	71.3
Fine and applied arts	8,371	77.2
Other special fields	1,495	77.7
Teacher training	1,334	85.8
Female profession (females 91–100%)		
Health and medicine	12,411	92.5
Total	57,709	58.0

(Source: Statistics Finland, 1993d, p. 104)

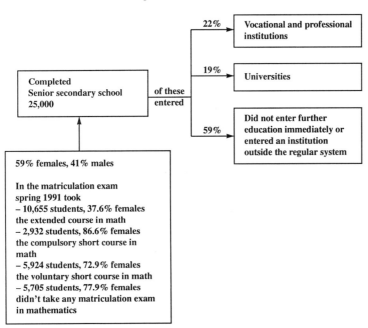

Figure 5: Choices made by those students who matriculated in spring 1991

(Sources: Rosenberg, 1992, pp. 18–20; Statistics Finland, 1993d, pp. 85–86)

In 1991, 45 percent of 19-year olds completed courses in biology and geography in senior secondary school, but only 17 to 19 percent completed extended courses in mathematics, physics, and chemistry, and less than 10 percent of the same age group took the matriculation examination in physics and chemistry (Science Evaluation Group, 1992). Of all answers in physics in the matriculation examination in 1991, 14 percent were produced by girls (Arminen, 1992). In 1990, the percentage was 15 percent and in 1993, 13 percent (Arminen, 1991, 1994). According to the Second International Science Study (IEA/SISS) in 1983/84, boys got better results than girls in biology, chemistry, and physics in Finland (Leimu, 1993).

More boys than girls choose the extended course in mathematics in the senior secondary school and proportionally more boys than girls get the highest marks in the matriculation exam in extensive mathematics. When making a chi-square test on the differences, the p-value is less than .0001 (see Table 3). The Leikola Committee pointed out that those who have studied the short course in mathematics but fail in the matriculation exam or do not try to take it at all must be considered as a problem. In this group of about 9,000 students, about 7,000 are girls (Kommittébetänkande, 1988).

Table 3: Differences between girls and boys in the matriculation examination in extended mathematics in 1986-1993 by mark
(1 the highest, i failed)

mark	1986	1987	1988	1989	1990	1991	1992	1993
1	-12.2	- 8.7	- 9.9	- 6.7	- 9.5	- 9.7	- 5.5	- 8.8
m	+1.0	+1.0	+2.5	+0.8	+2.2	- 0.1	+0.5	- 0.2
c	+3.5	+5.1	+2.8	+3.3	+3.5	+5.0	+0.9	+0.8
b	+3.8	+2.3	+1.7	+3.7	+2.4	+3.1	+1.6	+3.3
a	+1.7	+0.2	+2.4	+0.2	+1.8	+0.2	+2.0	+3.2
failed								
i	+2.2	+0.1	+0.5	- 1.3	- 0.4	+1.5	+0.5	+1.7

Note: In percentage points (proportion of girls in percent minus proportion of boys in per cent). Chi-square test gave p<.0001.

(Source: Rosenberg, 1986, 1987, 1988, 1989, 1991, 1992, 1993a, 1993b)

When looking at these data, one has to remember that the girls choosing extended mathematics belong to the elite of comprehensive school students with respect to learning and motivation for mathematics (Yrjönsuuri, 1989). As well, in 1971 as in 1992, 41 percent of all those taking the matriculation exam in mathematics had taken the extended course (Statistics Finland, 1993d). Between 1986 and 1993, the number of students taking the degree in mathematics in the matriculation examination has been quite stable (see Figure 6).

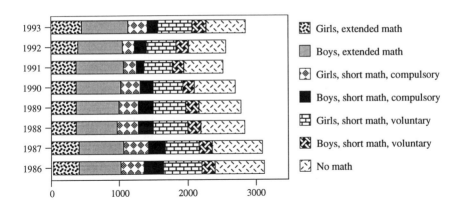

Figure 6: Number of students taking the matriculation in mathematics
and not taking it by year, type of exam and sex

(*Source: Rosenberg, 1986, 1987, 1988, 1989, 1991, 1992, 1993a, 1993b*)

As the educational fields of mathematics, natural sciences, and engineering in the universities require a thorough understanding of mathematics and natural sciences, many girls exclude themselves from these fields by not taking extended mathematics and physics in senior secondary school (Nurmi, 1993). Extensive courses in mathematics and physics are more often used as criteria for further education than are courses in optional languages (Lahelma, 1993). According to the Leikola Committee, about 40 percent of all new university students study fields where knowledge of mathematics and natural sciences is central, and more than 50 percent will go on to study mathematics and natural sciences to some degree (Kommittébetänkande, 1988).

The Finnish senior secondary school has been criticized for having too many compulsory courses in too many subjects (e.g., Science Evaluation Group, 1992). In the future there will be less compulsory courses and more room for choice.

I find the gender differences in senior secondary school depressing, not only because so few girls choose extensive mathematics but also because so many girls do not take the matriculation examination in mathematics at all. Those who fail in the matriculation examination or do not take it at all usually have weak mathematical knowledge (Kommittébetänkande, 1988). The fact that proportionally fewer girls than boys get the highest marks in the matriculation examination in extensive mathematics should be investigated. The concern for girls' choices in the senior secondary school should not be restricted to mathematics since kindred subjects, like physics and chemistry, are just as important when choosing what to study after senior secondary school.

THE UNIVERSITIES

The Finnish universities are gender-segregated as well: male students prefer engineering and female students humanities and community service (see Figure 3) (Nurmi, 1993).

In 1992, women took 54.7 percent of all master's degrees and 37.7 percent of all postgraduate degrees. In engineering, 13.6 percent of the postgraduate degrees and 11.8 percent of the doctoral degrees were taken by women. In mathematics and the natural sciences, the respective percentages were 31.3 and 33.6. Of the doctoral degrees in medicine, 34.6 percent were taken by women (Statistics Finland, 1993e).

When looking closer at the field of mathematics and natural sciences, you will find that women prefer biology, chemistry and biochemistry, and geography over mathematics, computer science, and physics (Hassi, 1986; Statistics Finland, 1993a). In 1992, at the University of Helsinki, almost 60 percent of the women studying in the Faculty of Mathematics and Natural Sciences studied biology, pharmacy, chemistry, biochemistry, and geology. The corresponding percentage for men was about 15 percent (see Figure 7).

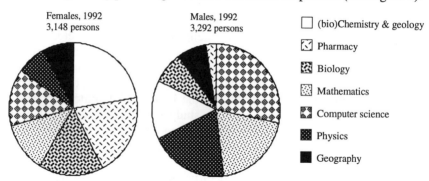

Figure 7: Proportion of women and men studying the different
fields of education at the faculty of mathematics and
natural sciences at the University of Helsinki in 1992

(Source: Helander, 1993, p. 29)

In 1988, about 36 percent of all degrees in mathematics awarded by universities in Finland had been taken by women (see Table 4). About 12 percent of the postgraduate degrees, licentiate and doctorate, had been taken by women. The professional field of education and research is by far the most popular among women.

At the end of 1992, about 36 percent of the master's degrees in mathematics had been taken by women (Statistics Finland, personal communication, April 1994). If you combine computer science and mathematics, there were 26 percent female students in 1989 (Statistics Finland, 1991). In math-

Table 4: Professional distribution in mathematics by sex and degree
in 1988, degrees awarded by all universities.

Degree	Education & research	Technical & business	Finance & insurance	Manufact.	Other	Total
BSc	1,526	489	254	186	688	3,143
female	718	162	123	47	265	41.8%
MSc	1,719	654	456	248	809	3,886
female	692	185	141	62	223	33.5%
PhLic	143	11	6	5	23	188
female	18	1	0	0	6	13.3%
PhD	180	2	4	5	14	205
female	7	0	1	0	3	5.4%
All	3,568	1,156	720	444	1,534	7,422
female	1,435	348	265	109	497	35.7%

(Source: Pylvänäinen, 1992, p. 104)

ematics and computer science programs at the third level, the percentage of
females has decreased from 31.3 percent in 1984 to 22.0 percent in 1991.
Agricultural, forestry, and fishery programs are the only other field showing
a decrease, from 42.9 to 39.2 percent females, during the same period
(Yearbook of Nordic Statistics, 1986, 1993). I find the decrease of females
in the mathematics and computer science programs quite disturbing.

The University of Helsinki is the largest university in Finland, with 6,044
new students in 1991 (Statistics Finland, 1993d), and is responsible for one-
third of the education given in science (Pylvänäinen, 1992). In the autumn
of 1992 in the Department of Mathematics, there were 1,156 students, of
whom 414, or about 36 percent, were female (about 33% of the new students were female). In 1992, 25 master's degrees were taken, 7 by women; 6
licentiate degrees, all taken by men; and 3 doctoral degrees, 1 by a woman.
In 1992, in the Faculty of Mathematics and Natural Sciences, 53.5 percent
of the master's degrees, 38.3 percent of the licentiate degrees, and 32.7 percent of the doctoral degrees were taken by women (Helander, 1993).

In 1991, at Åbo Akademi, the Swedish-language university, of the total
3,967 students, 59.2 percent of whom were female, 58 were studying in the
Department of Mathematics; 50 percent of these were female. In 1992, the
percentage of female students in the Department of Mathematics was 39.4.
In 1991, in the Faculty of Mathematics and Natural Sciences, there were a
total of 649 students, 51.3 percent of whom were female. In 1991 in this
faculty, 42 master's degrees were taken (52% by women); 3 licentiate
degrees (all by men), and 11 doctoral degrees (18% by women) (University

Secretary of Studies, personal communication, autumn 1993).

Keskitalo and Korhonen have found that the experience of the subject, whether or not it is interesting, is more important for girls' choices than for boys' (1984). Härkönen (1992) found in a survey concerning females studying engineering that the interest in mathematics had been important for choosing the field of engineering. Lagerspetz found that interest in the subject was more important for those studying in the Faculty of Mathematics and Natural Sciences than for those studying engineering. Both male and female students in engineering had a more instrumental attitude to life than students at the Faculty of Mathematics and Natural Sciences (1990). Liljander found that men, significantly more than women, saw studying as a hobby to develop personality (1991). In a study of 1985 to 1992 graduates from the Faculty of Mathematics and Natural Sciences at the University of Turku, Jutila found that women were more oriented towards a profession than men (Jutila, 1993).

The duration of study in Finland is too long. For example, the median age for a doctoral degree in mathematics and natural sciences is 34.5 years (Statistics Finland, 1993e). The universities are trying to solve this problem. Most women studying mathematics at the university will become teachers and few will continue their studies in mathematics at a postgraduate level.

I also want to point out that a new gender-segregated field of education, computer science, is emerging in Finland.

CONCLUSIONS

The policy in Finland has been to create equal opportunities for all. The reproduction cycle is, however, strong. The educational field at all levels and the labor market is gender segregated.

One of the problems in mathematics education is that relatively fewer girls choose the extended course in mathematics in the senior secondary school. As extensive mathematics and physics is a requirement for many of the fields of education popular among girls, for example, medicine, dentistry, and veterinary studies, there are not many left for the field mathematics. And a big proportion of those choosing mathematics at the university will probably become teachers.

Another cause for worry is why proportionally fewer girls than boys achieve the highest marks in the matriculation examination in extensive mathematics. Not having an answer, I will refrain from speculation.

In order to get more women into postgraduate studies in mathematics at the universities, we will probably have to start with the compulsory schools. The different patterns of task-choosing behavior by girls and boys, math self-concept, and attitudes towards mathematics and learning mathematics should be investigated. And the reason, or reasons, for the achievement gap in mathematics between boys and girls in senior secondary school must be explained.

NOTE

I want to express my gratitude to professor Ole Björkqvist at the Department of Teacher Education at Åbo Akademi and my parents for their support, and to Thomas Finne, Gunilla Myrskog, and Mikael Sundholm for helpful comments. I also want to thank the Department of Mathematics at Åbo Akademi for a solid and unbiased education in mathematics.

REFERENCES

Aitta, U. (1988). *Miesten ja naisten tasa-arvon toteutuminen akavalaisessa työelämässa* [The realization of gender equity in working life of members of Akava]. Helsinki, Finland: Akava.

Antikainen, A. (1983). *Remarks on the equality of opportunity in the Finnish education system* (Kasvatustieteiden osaston selosteita ja tiedotteita No. 39). Joensuu, Finland: Joensuun korkeakoulu, Kasvatustieteiden osasto.

Anttalainen, M-L. (1986). *Sukupuolen mukaan kahtiajakautuneet työmarkkinat pohjoismaissa* [The gendersegregated labour market in the Nordic countries] (Naistutkimusmonisteita No. 1). Helsinki, Finland: Tasa-arvoasiain neuvottelukunta.

Arminen, E. (1991). Fysiikan yo-koe K 1990 [The matriculation exam in physics spring 1990]. *Dimensio, 55*(2), 48–51.

Arminen, E. (1992). Fysiikan yo-koe K 1991 [The matriculation exam in physics spring 1991]. *Dimensio, 56*(2), 38–43.

Arminen, E. (1994). Fysiikan yo-koe K 1993 [The matriculation exam in physics spring 1993]. *Dimensio, 58*(1), 16–19.

Backman, J. (1994, February 2). Barnträdgårdslärarinstitut redo bli del av PF 1995 [Institute for preprimary teachers prepared to be part of PF 1995]. *Meddelanden från Åbo Akademi*, pp. 10–11.

Bergman, S. (1985). *Kvinnan och vetenskapssamfundet* [Women and the academy of science] (Kvinnoforskningsstenciler No. 1). Helsingfors, Finland: Delegationen för jämlikhetsärenden i Finland.

Björkqvist, O. (1994). *Utvärdering av matematikkunskaperna i årskurs 9 i grundskolan* [Assessment of mathematical knowledge in grade 9 in comprehensive school]. (Publikationer från pedagogiska fakulteten vid åbo Akademi No. 8). Vasa, Finland: Pedagogiska fakulteten, Institutionen för lärarutbildning.

Carlson, R. V. (1974). *Reform of secondary education in Finland* (Report No. 19). Oulu, Finland: University of Oulu, Institute of Behavioral Sciences.

Dimensio. (1989). Peruskoulun parhaat matemaatikot 1989 [Best mathematicians in comprehensive school 1989]. *Dimensio, 53*(4), 30.

Dimensio. (1994a). Lukion matematiikkakilpailu 1993 [The competition in senior secondary school 1993]. *Dimensio, 58*(1), 56–58.

Dimensio. (1994b). Peruskoulun matematiikkakilpailu 1993 [The competition in comprehensive school 1993]. *Dimensio, 58*(1), 58–59.

Haapasalo, L. (1992). *Murtolukukäsitteen konstruktivistinen oppiminen* [Constructivist learning of fractions] (Kasvatustieteiden tutkimuslaitoksen julkaisusarja A No. 51). Jyväskylä, Finland: Jyväskylän yliopisto, Kasvatustieteiden tutkimuslaitos.

174　LENA M. FINNE

Haapasalo, L. (1993). *Matematiikan opetussuunnitelmien lähtökohtia ja kehittämisnäkymiä* [Points of departure and development prospects in mathematics education]. Jyväskylä, Finland:Jyväskylän yliopisto, opettajankoulutuslaitos.

Haataja, E. (1989). Naisen tie kodista kouluun ja julkisuuteen [The road of women from home to school and in public]. In A. Haataja, E. Lahelma, & M. Saarnivaara (Eds.), *Se pieni ero* [That small difference] (pp. 21–58). Helsinki, Finland: Valtion painatuskeskus.

Hakaste, S. (1988). Male and female teachers working together in the secondary schools of Finland from the 1780s to the 1860s. In S. Seppo (Ed.), *The social role and evolution of the teaching profession in historical context.* (Conference papers for the 10th session of the International Standing Conference for the History of Education, vol. V, pp. 133–142). Joensuu, Finland: University of Joensuu, Faculty of Education.

Hakaste, S. (1992). *Yhteiskasvatuksen kehitys 1800-luvun Suomessa sekä vastaavia kehityslinjoja naapurimaissa* [The development of co-education in Finland during the 19th century and parallels in neighboring countries] (Research Report, No. 133). Helsinki, Finland: Helsingin yliopisto, kasvatustieteen laitos.

Härkönen, T. (1992). Naisteekareiden kokemuksia opiskelusta ja opiskelijaelämästä [The female engineering students' experiences of studying and student life]. Helsinki, Finland: Tekniikan akateemisten liitto.

Hassi, S. (1986). *Naiset ja tekniikka* [Women and engineering] (Tasa-arvoasiain neuvottelukunnan monisteita No. 6). Helsinki, Finland: Tasa-arvoasiain neuvottelukunta.

Helander, P. (Ed.). (1993). *Helsingin yliopisto. Tilastot* [The University of Helsinki. Statistics]. Helsinki, Finland: Helsingin Yliopisto.

Husu, L. (1991). Women's studies in Finland. In S. Bergman (Ed.), *Women's studies and research on women in the nordic countries* (pp. 21–30). Turku, Finland: Painotalo Gillot.

Jutila, H. (1993). *Turun yliopiston matemaattis-luonnontieteellisestä tiedekunnasta vuosina 1985-1992 valmistuneiden työllisyystilanne valmistumishetkellä* [The employment rate of the graduated students from the faculty of mathematics and natural sciences at the university of Turku in 1985-1992] (Hallintoviran julkaisusarja, No. 5). Turku, Finland; Turun yliopisto, opintoasiaintoimisto.

Kaipainen, U. (1994). *Koulutuksen kysyntä 1992* [The request for education 1992]. (Muistio No. 137). Helsinki, Finland: Official Statistics of Finland.

Kangasniemi, E. (1989). *Opetussuunnitelma ja matematiikan koulusaavutukset* [Curriculum and student achievement in mathematics] (Kasvatustieteiden tutkimuslaitoksen julkaisusarja A No. 28). Jyväskylä, Finland: Kasvatustieteiden tutkimuslaitos.

Karjalainen, O. (1982). *Matematiikan opetuksen affektiiviset tavoitteet ja niiden totutumisen arvioiminen peruskoulun ala-asteella* [The affective objectives of mathematics teaching and an evaluation of their attainment in the lower comprehensive school] (Reports from the Faculty of Education No. 10). Oulu, Finland: Oulun yliopisto, Kasvatustieteiden tiedekunta.

Keskitalo, M., & Korhonen, R. (1984). *Korkeakouluihin valikoituminen ja ylioppilasvalmennuskurssit yhteiskunnan, korkeakoulun ja yksilön kautta* [University preparatory courses and getting into universities form the point of view of society, university and the individual] (Oulun yliopiston kasvatustieteiden tiedekunnan tutkimuksia No. 26). Oulu, Finland: Oulun yliopisto, Kasvatustieteiden tiedekunta.

Kivikkokangas-Sandgren, R. (1989). *First women geographers in Finland 1890–1930. A development perspective* (Publicationes instituti Geographici univeristatis Helsingiensis, No. C4). Helsinki, Finland: Helsingin yliopisto, maantieteen laitos.

Klinge, M. (1970). Yliopiston vaiheet [The different periods in the university]. In N. Luukanen (Ed.), *Helsingin yliopisto — historiaa ja nykypäivää* [The university of Helsinki — history and today] (pp. 9–33). Porvoo-Helsinki, Finland: Werner Söderström.

Kommittébetänkande (1982). *Naisten tutkijanuran ongelmat ja esteet — Problem och hinder i karriären för kvinnliga forskare* [Problems and obstacles in the academical career of women]. (Kommittébetänkande No. 33). Helsinki, Finland: Valtion painatuskeskus.

Kommittébetänkande (1988). *Matemaattis-luonnontieteellisen perussivistyksen komitean välimietintö* [Intermediate report of the committee of general knowledge in mathematics and natural sciences]. (Kommiteamietintö, No. 33). Helsinki, Finland: Valtion painatuskeskus.

Kommittébetänkande (1989). *Matemaattis-luonnontieteellisen perussivistyksen komitean välimietintö* [Final report of the Committee of General Knowledge in Mathematics and Natural Sciences]. (Kommiteamietintö, No. 45). Helsinki, Finland: Valtion painatuskeskus.

Korhonen, H. (1994). Peruskoulun valtakunnallinen 9. luokan matematiikan koe 1993 [The national test in mathematics in the ninth grade in comprehensive school]. *Dimensio, 58*(3), 56–58.

Kupari, P. (1983a). *Millaista matematiikkaa peruskoulun päättyessä osataan?* [What kind of mathematics do students master at the end of comprehensive school?] (Kasvatustieteiden tutkimuslaitoksen julkaisuja No. 342). Jyväskylä, Finland: Jyväskylän yliopisto, Kasvatustieteiden tutkimuslaitos.

Kupari, P. (1983b). Tytöt, pojat ja matikkapää: Sukupuoli ja matematiikka. [Girls, boys and a head for mathematics: Gender and mathematics]. In H. Varsa (Ed.), *Naiset, tekniikka ja luonnontieteet* [Women, technology and natural sciences] (pp. 81–91). (Tasa-arvoasiain neuvottelukunnan monisteita No. 8). Helsinki, Finland: Valtion painatuskeskus.

Kupari, P. (1993). Matematiken i den finska grundskolan. Attityder och kunskaper [Mathematics in the Finnish comprehensive school. Attitudes and achievements]. *Nordisk matematikkdidaktikk, 1*(2), 30–58.

Lagerspetz, O. (1990). *Kvinnor och män i teknikens värld* [Women and men in the world of technology]. (Forskningsrapport No. 6). Åbo, Finland: Åbo Akademi, Institutet för kvinnoforskning.

Lahelma, E. (1987). *Sukupuolten tasa-arvo koulussa* [The equity of gender at school] (Opetusministeriön suunnittelysihteeristön julkaisuja No. 4). Helsinki, Finland: Vation painatuskeskus.

Lahelma, E. (1989). Koulu vaikuttajaksi — mitä tehdään [Be influenced by school — what should be done]. In A. Haataja, E. Lahelma, & M. Saarnivaara (Eds.), *Se pieni ero* [That small difference] (pp. 21–58). Helsinki, Finland: Valtion painatuskeskus.

Lahelma, E. (1993). *Policies of gender and equal opportunities in curriculum development: discussing the situation in Finland in Britain.* Helsinki, Finland: University of Helsinki, Department of Education.

Latva, H. (1991). Peruskoulun matematiikkakilpailun loppukilpailu [The final competition of the competition in mathematics for comprehensive school]. *Dimensio, 55*(4), 64.

Lehtinen, M. & Saksman, E. (1992). Matematiikkaolympialaisen 1992 Moskovassa [The olympics in mathematics in 1992 in Moscow]. *Dimensio*, 56(9), 34–40.

Lehtinen, M. (1993). 34. Matematiikkaolympialaiset Istanbulissa [The 34th olympics in mathematics in Istanbul]. *Dimensio*, 57(7), 35–37.

Leimu, K. (1993). Suomen luonnontieteiden koulusaavutukset kansainvälisessä vertailussa [International comparison of Finnish schoolresults in natural sciences]. In Statistics Finland, *Koulutus* [Education], (pp. 249–260) (No. 7). Helsinki, Finland: Official Statistics of Finland.

Liljander, J-P. (1991). *Yliopistoinstituutio kulttuurisena uusintajana* [The university institution as an agent of cultural reproduction] (report, No. 40A). Jyväskylä, Finland: Jyväskylän yliopisto, kasvatustieteiden tutkimuslaitos.

Linnanmäki, K. (1990). *Matematiksvårigheter hos elever i årskurs 6 i tre skolor i Svensk-Finland* [Pupils' difficulties in mathematics in the sixth grade in three schools of the Swedish-language schools in Finland]. (Rapporter från pedagogiska fakulteten No. 30). Vasa, Finland: Pedagogiska fakulteten.

Lähteenmaa, J., & Siurala, L. (1992). *Youth and change*. Helsinki, Finland: Official Statistics of Finland.

Manninen, M. (1993). *Women in Finland*. Helsinki, Finland: The National Council of Women of Finland.

Martikainen, R. & Yli-Pieltilä, P. (1992). *Työehdot ja sukupuoli — sokeat sopimukset* [Collective agreements and gender — blind agreements] (sarja T-julkaisuja No. 12). Tampere, Finland: Tampereen yliopisto, Yhteiskuntatieteiden tutkimuslaitos, työelämän tutkimuskeskus.

Meri, M. (1992). *Miten piilo-opetussuunitelma toteutuu* [The ralization of the latent curriculum] (Research Report No. 104). Helsinki, Finland: Helsingin yliopisto, Kasvatustieteellinen tiedekunta.

Metso, T. (1992). Yhdessä vai erikseen? Tytöt ja piilo-opetussuunnitelma [Together or separate? Girls and the latent curriculum]. In S. Näre & J. Lähteenmaa (Ed.), *Letit heilumaan* [Wave your pigtails] (pp. 271–283). Helsinki, Finland: Suomalaisen kirjallisuuden seura.

Ministry of Education (1992). *Developments in education 1990–1992 Finland.* (Reference Publications No. 16). Helsinki, Finland: Valtion painatuskeskus.

Mäkelä, R. (1991) Sara Wacklin tyttöjen koulutksen uranuurtajana [Sara Wacklin as predecessor for education for girls]. *Kasvatus*, 22, 227–232.

Nummenmaa, A. R. (1989). Toisen sukupuolen ammatti nuoren koulutusvalintana [To choose the profession of the other gender]. *Kasvatus*, 20, 494–503.

Nummenmaa, A. R. (1990). Miesten ammatti tytön koulutusvalinta [Girl's choice of male job]. *Naistutkimus*, 1, 20–34.

Nurmi, S. (1993). Korkeakoulutuksen pyrkijäsumat eri koulutusaloilla [The application stream to higher education in different fields of education]. *Kasvatus*, 24, 287–295.

Ojala, M. (1993). Suomalainen esiopetus kansainvälisessä vertailussa [International comparison of Finnish preprimary education]. In Statistics Finland (1993b), *Koulutus* [Education] (pp. 243–247) (Koulutus No. 7). Helsinki, Finland: Official Statistics of Finland.

Onikki. E. & Ranta, H. (1992). *Suomen laki II* [Finnish law II]. Helsinki, Finland: Lakimiesliiton kustannus.

Opetusministeriö. (1993). *KOTA 1992.* Helsinki, Finland: Opetusministeriö, Korkeakoulu- ja tideosasto.

Pehkonen, E. (1992). *Problem fields in mathematics teaching* (Research Report No. 108). Helsinki, Finland: University of Helsinki, Department of Teacher Education.

Pehkonen, E. (1994). Seventh-graders' experiences and wishes about mathematics teaching in Finland. *Nordisk matematikk didaktikk, 2*(1), 31–45.

Pylvänäinen, M. (1992). *Statistics on higher education in mathematics and natural science in Finland 1971–1990.* (Publications from the Council of Higher Education No. 4). Jyväskylä, Finland: Ministry of Education, Science Evaluation Group.

Raivola, R. (1989). *Opettajan ammatin historia* [The history of the profession of teachers] (Tutkimusraportti No. A44). Tampere, Finland: Tampereen yliopisto, Kasvatustieteen laitos.

Rantalaiho, L. (1993). Gender system of Finnish society. In L. Rantalaiho (Ed.), *Social changes and the status of women: The experience of Finland and the USSR* (pp. 2–8) (Working papers No. 3). Tampere, Finland: University of Tampere, Research Institute for Social Studies.

Rosenberg, E. (1986). Matematiikan koe K 1986 [The test in mathematics spring 1986]. *Dimensio, 50*(9), 8–17.

Rosenberg, E. (1987). Matematiikan koe K 1987 [The test in mathematics spring 1987]. *Dimensio, 51*(9), 14–22.

Rosenberg, E. (1988). Matematiikan koe K 1988 [The test in mathematics spring 1988]. *Dimensio, 52*(9), 16–25.

Rosenberg, E. (1989). Matematiikan koe K 1989 [The test in mathematics spring 1989]. *Dimensio, 53*(9), 38–45.

Rosenberg, E. (1991). Matematiikan koe K 1990 [The test in mathematics spring 1990]. *Dimensio, 55*(2), 38–47.

Rosenberg, E. (1992). Matematiikan koe K 1991 [The test in mathematics spring 1991]. *Dimensio, 56*(2), 14–22, 27.

Rosenberg, E. (1993a). Matematiikan koe K 1992 [The test in mathematics spring 1992]. *Dimensio, 57*(2), 10–25.

Rosenberg, E. (1993b). Matematiikan koe K 1993 [The test in mathematics spring 1993]. *Dimensio, 57*(8–9), 35–44.

Saranen, E. (1993). Lukion yleisen oppimäärän opiskelijoiden matematiikan taidot ja käsitykset matematiikasta, osa I [Students studying short mathematics in senior secondary school: Their knowledge and conceptions of mathematics]. *Dimensio, 57*(1), 14–18.

Science Evalutation Group. (1992). *Report on the external evaluation of higher education in mathematics and natural science in Finland.* (Korkeakouluneuvoston julkaisujan No. 3). Jyväskylä, Finland: Ministry of Education.

Seppänen, R., Toivari, A-L., Pippola, L. & Norlamo, P. (1992). Matematiikkakilpailun voitot Kehä III:n ulkopuolelle [The prizes in the competition in mathematics went outside Ring III]. *Dimensio, 56*(3), 30,35.

Statistics Finland. (1980). *Postition of women* (Statistical Surveys No. 65). Helsinki, Finland: Official Statistics of Finland.

Statistics Finland. (1984). *Postition of women* (Statistical Surveys No. 72). Helsinki, Finland: Official Statistics of Finland.

Statistics Finland. (1991). *Koulutus* [Education in Finland 1991] (Koulutus ja tutkimus No. 11). Helsinki, Finland: Official Statistics of Finland.

Statistics Finland. (1993a). *Statistical yearbook of Finland 1993.* Helsinki, Finland:

178 LENA M. FINNE

Official Statistics of Finland.

Statistics Finland. (1993b). *Väestön koulutsrakenne kunnittain 31.12.1991* [The educational level of people by municipality 31.12.1991] (Koulutus No. 2). Helsinki, Finland: Official Statistics of Finland.

Statistics Finland. (1993c). *Lukiot syyslukukaudella 1993* [Senior secondary schools in autumn term 1993] (Koulutus No. 6). Helsinki, Finland: Official Statistics of Finland.

Statistics Finland. (1993d). *Koulutus* [Education]. (Koulutus No. 7) Helsinki, Finland: Official Statistics of Finland.

Statistics Finland. (1993e). *Korkeakoulut 1993* [Universities 1993] (Koulutus No. 10). Helsinki, Finland: Official Statistics of Finland.

Statistics Finland. (1993f). *Ammatilliset oppilaitokset 1993* [Vocational institutions 1993] (Koulutus No. 11). Helsinki Finland: Official Statistics of Finland.

Statistics Finland. (1993g). *Palkkatilasto 1992/1993* [Wages and salaries 1992/1993] (Palkat No. 20). Helsinki, Finland: Official Statistics of Finland.

Statistics Finland. (1994). *Väestön koulutustaso kunnittain 1991* [The educational level of people by municipality 1991] (Koulutus No. 1). Helsinki, Finland: Official Statistics of Finland.

Suomen valtiokalenteri. (1992). Helsinki, Finland: Helsingin yliopisto.

Sysiharju, A-L. (1984). *Women as educators: Employees of schools in Finland* (Research report No. 1984-22). Helsinki, Finland: University of Helsinki, Department of Teacher Education.

Tasa-arvo toimikunta (1990). *Sukupulten välisen tasa-arvon edistäminen Helsingin yliopistossa. Ehdotus tasa-arvosuunnitelmaksi* [To promote gender equity in the University of Helsinki. A suggestion to gender equity plan]. Helsinki, Finland: Helsingin yliopisto.

Tasa-arvolakiin ei naiskiintiöitä [No female quotas in the Act of equality]. (1994, May). *Satakunnan Kansa*, p. 1.

Tilastokeskus. (1975). *Statistical yearbook of Finland 1975.* Helsinki Finland: General Statistical Offices.

Toivari, A-L. (1990). Lukion matematiikkakilpailu 1989 [The competition in upper secondary school 1989]. *Dimensio, 54*(2), 58.

Työttömien määrä putosi alle 500 000:n [The unemployment rate is now less than 500,000]. (1994, May). *Satakunnan Kansa*, p. 14.

Utbildningsstyrelsen. (1994). *Grunderna för grundskolans läroplan* [The basis of the curriculum for comprehensive school]. Helsingfors, 1994: Utbildningsstyrelsen.

Wilkama, S. (1938). *Naissivistyksen periaatteiden kehitys Suomessa 1840-1880-luvuilla* [The evolution of the principles of women education in Finland 1840-1880]. Helsinki, Finland: Suomalaisen kirjallisuuden seuran kirjapainon.

Yearbook of Nordic Statistics. (1986). Nordic Council of Ministers.

Yearbook of Nordic Statistics. (1993). Nordic Council of Ministers.

Yrjönsuuri, R. (1989). *Lukiolaisten opiskeluorientaatiot ja menestyminen matematiikassa* [Secondary school students' study of orientations and achievements in mathematics] (Tutkimusraportit No. 120). Helsinki, Finland: Helsingin yliopisto, Kasvatustieteiden laitos.

CORNELIA NIEDERDRENK–FELGNER

GENDER AND MATHEMATICS EDUCATION: A GERMAN VIEW

Up to the middle of the 19th century, education for girls and for boys meant two quite different things in Germany. While boys were prepared for their future vocations and professions, the destiny of a girl from the higher and middle classes was seen in being a good housewife and mother. Education for girls was, in general, considered neither necessary nor in the public interest, nor was any public responsibility felt for it. It was reserved for boys. Girls were only allowed to participate in order to learn the fundamental skills of writing, reading, and calculating.

After 1820, however, more and more private schools for girls came into existence, the so-called *höhere Töchterschulen* (girl's *lyceum*/high school for girls). They prepared girls for their future role in the family, but also for those jobs which became available to middle-class women in the changing industrial society of the late 19th century. This was probably the reason why the state accepted responsibility for the education of girls and established a special school system for them. At this point, clarification as to what sort of formal education girls should receive became unavoidable. It was undisputed in those days that girls and women were very emotional human beings, and both the methods employed and the subject matter taught in schools were chosen accordingly.

Women and girls were considered less capable of abstract thinking and as not having the same intellectual abilities as men. Curricula in those days pointed out that it would be inappropriate to demand too much of girls by having them sit too long or forcing them to exert their minds too much.

It was particularly this prejudice — the assumption that there was such a thing as a gender difference in mental ability — that had an unfavorable effect on mathematical education and on curricula for girls' schools. Until the beginning of the 20th century, it was true to say that girls did not know mathematics, simply because they had no chance of learning it at school. They were only instructed in fundamental calculation, mainly for domestic applications.

At the same time, about 1850, there was an elaborate discussion about the aims of mathematical education for boys. One group regarded mathematics as a form of cultivation of the mind; others pointed out that the progress of technology required mathematical thinking. No one disputed that boys were able to do whatever abstract thinking was expected of them.

From the beginning of the 20th century, at the latest, mathematics had become a recognized and highly valued subject at boys' schools, to which

G. Hanna (ed.), Towards Gender Equity in Mathematics Education, 179–195.
© 1996 *Kluwer Academic Publishers. Printed in the Netherlands.*

great importance was attached. It was considered indispensable for the *Abitur*, the final school examination which constitutes the qualification for university study.

The admission of girls to the *Abitur* was the reason why in 1908 mathematics was, for the first time, introduced into the curricula of girls' schools. But it was taken for granted that differences should exist between girls' and boys' schools regarding the content and number of lessons in this subject. This was not changed until a reform of the educational system in 1924. It was only then that the aim of equal education for girls and boys was defined — within separate single-sex school systems, of course.

The Nazi era in Germany meant a setback. Nazi ideology once more defined the mother role as the aim of girls' education. The number of mathematics lessons per week in girls' schools was cut down. After World War II, school systems in the newly emerging states (*Länder*) that formed the Federal Republic — in West Germany, in other words — once more resembled the situation in the pre-Nazi Weimar Republic. In the (West) German system, the four-year comprehensive primary school (*Grundschule*) is followed by a three-tier system of secondary education. The different secondary school types are:

- *Hauptschule* — the "standard" type of secondary school although today only attended by a minority of the pupils
- *Realschule* — a school attended for six years, up to age 16, with a technical/vocational bias
- *Gymnasium* — the traditional secondary school leading to higher education, like the British grammar school, but attended for nine years

The *Gymnasium* was divided into different categories where special emphasis was laid on classical languages, modern languages, and mathematics and science. Initially there were separate single-sex schools of every category. It was only in the 1970s that mixed (co-educational) secondary schools of all types became the rule in all states of West Germany. Curricula for boys and girls were more or less the same, but a much higher proportion of the girls who went to a *Gymnasium* — more than 60 percent — went to the modern-language type, whereas 35 percent of the boys went to the modern-language and the mathematics/science type respectively. This tendency continued in the 1970s when a reform was introduced for the pre-*Abitur* upper grades. Now from the 11th grade upwards, pupils have a choice in various subjects between a Standard Course (*Grundkurs*) and an Advanced Course with a more advanced curriculum and more lessons per week (somewhat misleadingly called "Achievement Course"— *Leistungskurs* in German).

Bettina Srocke has published a more detailed account of the development of mathematics as a teaching subject, taking into account the different conditions for boys and girls (Srocke, 1989). In conclusion of my own summary of the history of the problem, I would like to emphasize two points that seem crucial.

Women were said to be incapable of abstract thinking and of learning mathematics long before they had the possibility of doing it. This prejudice was not based on any relevant experience, and the structural and formal inequality between women and men was never considered. Women were treated and educated in what was considered an "appropriate" way, with the result that they actually corresponded to the lack of independence and rationality that was attributed to them.

It was formal criteria like university entrance requirements that finally gave women access to the same mathematical education as men. From then on (at least) official policy was no longer based on consideration of different mental abilities between genders. But no attention was paid to women's interests. There was no questioning of the fact that curricula and teaching methods that had been *developed by men for boys* were used for girls as well.

WOMEN/GIRLS AND MATHEMATICS — SOME DATA

If figures give us facts about the proportion of girls and women in different sectors of the educational system, the facts are clear enough. Mathematics is still a male domain in Germany.

Let us first look at some data from the upper grades of *Gymnasium*. Starting from Grade 11 or 12, students may opt for a mathematics curriculum, either of the Standard or Advanced Course. In North Rhine Westphalia, one of the most highly populated German states, girls and boys in Grades 11, 12, and 13 chose these courses in the school year 1992/93 as shown in Table 1:

Table 1: 1992/93 Enrolments in grades 11–13 mathematics courses, North Rhine, Westphalia

	Girls		Boys	
Mathematics	numbers	%	numbers	%
Standard course	61,250	55.4	49,293	44.6
Advanced course	10,439	35.8	18,733	64.2

(Source: Landesamt für Datenverarbeitung und Statistik NRW, 1993)

Similar figures come from Bavaria (Beerman, Heller, & Menacher, 1992, p. 18). We cannot provide statistics for the situation in the whole of the Federal Republic of Germany, as the officially available figures from some states — Baden-Württemberg, for example — do not differentiate between genders, a fact which may be a clue to the importance attached to issues like gender and mathematics education by the authorities. We will come back to this later.

At German universities there is a further decrease in the proportion of women, to even a dramatic extent once we take senior teaching staff into consideration. Figure 1 shows the proportion of women at different levels, for all university subjects and for mathematics.

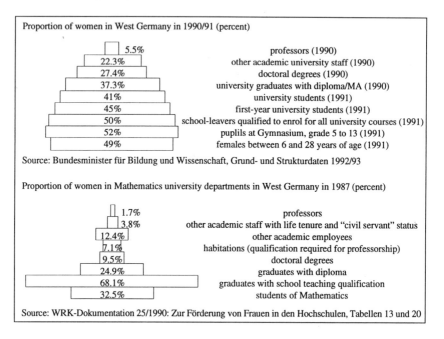

Figure 1

Back to the secondary school again. What is the proportion of women school teachers? Figure 2 illustrates the statistics for mathematics teachers at state schools in Baden-Württemberg in September 1991, which may serve as an example.

The figures show that significantly fewer girls and young women than boys and young men chose mathematics as one of their special subjects at school or as a university course. At the level of primary schools, there are more women teachers than men teachers, but the proportion is different if we consider the number of lessons taught per week. A considerable number of women teachers are only employed part-time.

In Baden-Württemberg, only 20 percent of the teaching staff of the *Gymnasien* are women, and of these only one third work full-time — statistics that are not representative for the whole of the Federal Republic, but a very blatant example of a general tendency that exists everywhere.

At the university level, the percentage of women teachers is marginal. The same is true (to a slightly lesser extent) if we look at staff in academic institutions — that is, universities, teacher training colleges, and research

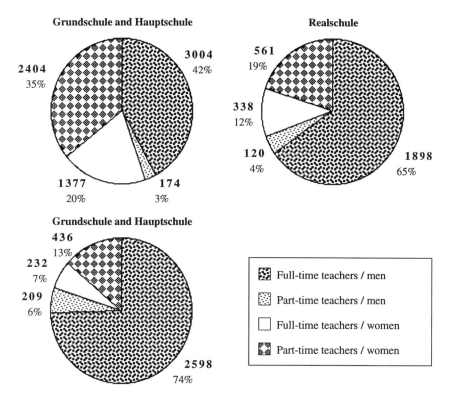

Figure 2

(*Source: Statistisches Landesamt Baden-Württemberg*)

institutes — dealing with mathematics education. A members' directory for 1992 of mathematics educators in Germany and Austria shows that 14 percent of these were women.

At first glance, one might conclude that girls do not take advantage of the equal opportunities in education now offered to them (after long battles). A closer look at mathematics as a school subject will give plausibility to some different explanations which I will examine later in this article.

RESEARCH ON GENDER AND MATHEMATICS

Attitudes Towards Mathematics

A number of studies and research projects in Germany in recent years have dealt with the causes of the different attitudes towards mathematics of girls/women and boys/men respectively. These different attitudes not only

find expression in the data that have been cited, but also in a variety of prejudices and gender-specific attributions and assumptions in daily life.

Usually a broader range of questions was dealt with in this research, covering the whole area of mathematics, science, and technology. One example
is the survey by Ilse Brehmer, Hildegard Küllchen, and Lisa Sommer
(Brehmer, Küllchen & Sommer, 1989). They studied the conditions which
determine the choice of subjects in pre-*Abitur* upper grades. They conducted interviews with 72 pupils from 12 *Gymnasium*-type schools in North
Rhine Westphalia. Results of these interviews clearly reveal certain gender-
specific differences.

When pupils were asked why they chose Advanced Courses, it was significant that girls in general, and those girls who were interested in mathematics and science courses in particular, mentioned less frequently that they
wished to study the subject at university and start a career in it. For boys,
on the other hand, it was a matter of course to point out that their subjects
were useful for a specific career. To the researchers this indicated a striking
lack of career-focussed women who could serve as role models and an
anticipation by the girls of their chances of (not) finding suitable jobs in the
field of mathematics and science.

Success at school turned out to be one of the key motivations for choosing an Advanced Course. But boys and girls gave different interpretations
to what success in a school course meant for their performance and ability
in general. It was boys rather than girls with good grades who tended to
have a more stable and positive concept of their own ability. Various interviews showed girls with good grades to be self-critical, whereas boys tended to view themselves too positively.

Gender-specific behavior was regarded as irrelevant by the pupils. Girls
even seemed to evade discussion about it in the context of everyday school
life.

A study by Gabriele Kaiser-Messmer deals more specifically with attitudes to mathematics as a school subject (Kaiser-Messmer, 1993). In
1988/89, 748 pupils of different ages and school types were interviewed
about their interest in mathematics. The author comes to the conclusion that
girls make more open-minded statements than boys about mathematical
abilities and gender roles, but this does not correspond to the way a girl
views her own person and future role. Girls still tend to lack self-confidence regarding mathematical ability, and their plans for their lives tend to
remain within the scope of traditional ideas, where great importance is
attached to social acceptability.

Similar conclusions were reached in psychological research on the
development of young people's academic interests during adolescence. Bettina Hannover compared conditions under which girls in co-ed and single-
sex environments chose "unfeminine" school subjects (Hannover, 1992).
Her study is centred on what she calls a person's spontaneous self-concept
— a description of those aspects of a person's identity which are regarded

as non-traditional, new, or special in a given situation. According to these results, one's sex as part of one's self-concept is a factor that can prevent girls from developing an interest in subjects that are traditionally reserved for boys. As a result of the presence of males of the same age, this is more significant in co-ed than in single-sex schools or single-sex teaching groups. Separate single-sex teaching groups for boys and girls is, in fact, what the author proposes for mathematics and science subjects in order to minimize this negative influence.

All the studies show that young people's attitudes to different school subjects cannot simply be looked at in isolation as the special academic interests of individuals. Still decisive are traditional ideas in their environment about roles and gender-specific attributions: mathematics, science, and technology subjects tend to be linked with the masculine domain, languages and social sciences with the feminine domain. These traditional ideas are promoted in all forms of education from early childhood throughout one's school years to career guidance and job advertisements, exercising a lasting influence on young people and their plans for their lives.

Ability, Achievement, Giftedness

With one or two exceptions, there seems to be little fundamental research on differences between boys and girls regarding mathematical performance and cognitive abilities. Usually, reference is made to relevant studies from the United States (e.g., Benbow & Stanley, 1980; Fennema, 1978; Hyde, Fennema, & Lamon, 1990; Linn & Hyde, 1989), pointing out that the results of this research cannot be easily transferred because of the great differences between the educational systems (Kaiser-Messmer, 1989).

Studies on gifted students in Germany more or less confirmed what similar studies had found in the United States. According to a study in Hamburg involving young people who were highly gifted in Mathematics and participated in the SAT-M, the proportion of boys in the leading group was significantly higher. In another test specially developed in Hamburg, where more spontaneity and creativity could be introduced by the subjects, girls were in the lead (Birx, 1988).

A study covering gifted young people of all ages between six and 18 showed that the academic achievement of girls — expressed in school marks — was higher in nearly all subjects. Boys' achievement was higher in mathematics and science subjects. The study also revealed that the girls' achievement — as compared with the boys' — became worse during adolescence (Heller, 1990).

These findings for gifted students are confirmed by the small proportion of girl participants in the Federal Mathematics Contest (*Bundeswettbewerb Mathematik*), a national program to discover and encourage young mathematical talents. Participation in this contest is one of the opportunities that exist in Germany to qualify for the International Mathematical Olympics (IMO).

Available figures for the years between 1970 and 1992 are given in Table 2.

Table 2: Participants in Federal Mathematics Contest, 1970–1992

| | Contest participants | | Award winners | |
	Boys	Girls	Boys	Girls
1st round	31 192	4 226 11.93%	16 532	1 574 8.69%
2nd round	6 219	479 7.15%	3 802	249 6.15%
3rd round	1 285	61 4.53%	356	16 4.30%

(Source: Bundeswettbewerb Mathematik)

An interesting cognitive science point of view has been introduced by Inge Schwank (Schwank, 1994). Her assumption is that there are two different types of mental structures which she calls *predicative* and *functional*. Predicative thinking emphasizes the preference for thinking in terms of relationships and judgments, leading to a static representation with clearly defined terms. Functional thinking emphasizes the thinking in terms of courses and modes of action, leading to a dynamic representation with a corresponding terminology. An empirical study based on assignments involving the formulation of mathematical terminology examined this theory of mental structures and looked at whether the subjects showed a stable preference for one or the other type of thinking and its corresponding behavior. One hypothesis of the experiment was that nearly all females' behavior will correspond to predicative thinking, while more male subjects will behave in accordance with functional than with predicative thinking.

These hypotheses were confirmed, but only on the basis of a very small sample involving 60 pupils aged 13 (7th grade) of whom 10 attended a *Realschule*, 32 a *Gymnasium* in Germany, and 18 a school in China (in order to obtain clues about possible cultural differences). Schwank herself regards her study as a contribution to women's studies that explains gender-specific differences in thinking and problem-solving in a way that is different from the causes that are usually found using social science methods. She does not regard such causes as exclusive explanations (Schwank, 1992, p. 69ff). On the basis of her findings, she endeavors to explain well-known gender-specific behavior patterns that are attributed to females (holding back, co-operating, looking for broader contexts) and males (playful approach, trying out first). The "soft" and "hard" programming styles defined by Sherry Turkle (Turkle, 1984) are also attributed to functional and predicative thinking.

What remains unclear are the consequences of these findings. One of Schwank's proposals is to increase the amount of "predicative" abstract concepts and theory in the teaching of mathematics, informatics, and physics, in order to appeal to girls (Schwank, 1994, p. 39). According to one of her previously published papers, compensating measures like training girls in functional thinking may be the appropriate answer (Schwank, 1992, p. 86). This may be (mis-?)understood as a "deficit" approach, even bordering on "biological" explanations for gender-specific cognitive differences.

A readable, knowledgeable, and up-to-date survey of the discussion on giftedness is contained in the book by Lilly Beerman, Kurt A. Heller, and Pauline Menacher (Beerman, Heller, & Menacher, 1992). It explains different theories and explanations from biology, psychology, and social psychology and presents the results of their research. This book is particularly helpful as a basic reader for teachers who are confronted with the (still) widespread gender-specific attribution of giftedness for specific subjects, especially mathematics and science.

Teacher/Student Interaction in the Teaching of Mathematics

In one of the first studies of gender-related behavior patterns and classroom interactions in Germany, fourth-grade primary school lessons in mathematics, project work, and German were monitored (Frasch & Wagner, 1982). This publication was the basis of what is now an established line of educational research. Its results have been confirmed repeatedly. The main result was that boys receive a significantly higher degree of attention from the teacher than girls do. Boys were more frequently called to answer questions, praised, reprimanded, and reminded of discipline. In mathematics lessons and project work, teachers more frequently began an interaction with boys than with girls and called boys rather than girls who had not raised their hands.

What gave a new impetus in this study was the fact that it considered for the first time as factors of gender-specific differences in schools not only the young people's individual qualities and abilities, plus outside influences, but also the behavior of the teaching staff. The researchers did not explain their results by citing more lively behavior on the part of boys to which teachers simply responded, but came to the conclusion that teachers were guided by subjective attitudes and role attributions in their regard of boys' contributions to a lesson as more valuable and worthy of their attention than girls' contributions. Support for boys is considered so important that teachers turn to them of their own accord, while girls have to ask for attention.

In the meantime, a variety of studies and research programs in Germany have dealt with classroom interaction, though not specifically with mathematics teaching. The specific situation of mathematics teaching is, however, the topic of an Austrian study which is relevant to the German situation as well. German mathematics lessons in secondary school are characterized by a form of instructional dialogue during which the content of the lessons is introduced through questions by the teacher, involving as many pupils as possible. Sometimes this style of conducting a lesson leads to a highly developed culture of communication, with high level conversations between teacher and pupils. There is, however, an inherent danger that this dialogue turns into a teacher-centred "question-and-answer game" with implicit rules that are known to all participants (Jungwirth, 1990). The "questions" are not genuine questions: the person who asks already knows the answers; specific answers are expected from the pupils, and these are frequently judged "right" or "wrong" in the teacher's feedback. The pupils play this game knowing that it is a game and not a genuine dialogue or conversation. In a case-study, Helga Jungwirth has examined gender-specific differences in responding to such a "small-step" style of instruction (Jungwirth, 1990). Her observations suggest that boys tend to handle the "rules of the game" better than girls, which means that they conform more easily to the teacher's expectations, which are, of course, geared to this style of instruction. The well-known fact that girls often withdraw from oral participation in mathematics lessons is not necessarily a sign of poor achievement. It could also indicate dissatisfaction with the style of the lesson, unreadiness to play the assigned part in this "instructional game."

School Textbooks and Educational Material

In German schools, school textbooks play an important role, especially in the teaching of mathematics. These books contain the syllabus for the school year, which is to a large extent prescribed, regulated, and defined. The books are usually in two sections: an instructional section which presents the learning items and a more lengthy section with problem-solving assignments.

Even those teachers who devise their own learning units usually take the assignments from the textbook, and schools choose those textbooks whose problem assignments appeal to the teaching staff.

Analyses of mathematics textbooks have shown that the texts of the assignment sections often perpetuate role stereotypes. This result emerged not only in previously published surveys (Dick, 1991; Glötzner, 1991; Lopatecki & Lücking, 1989) but also in a recent examination (Kaiser-Messmer, 1994) of three different series of mathematics textbooks. These books appeared between 1983 and 1990 with official sanction in the state of Hesse and are still in use. The study revealed that in these texts:

- females are under-represented in terms of number of mentions and appearances
- significantly fewer women are mentioned as being engaged in some sort of vocational activity; the general image is still one of a woman being part of a family and responsible for her household, while a man is doing his job

- men are shown as active, initiating, competent persons, while the women's range of activities and responsibilities appears more restricted (It is noteworthy, however, that textbook authors now appear to be more careful when boys and girls — that is, non-adults — appear in the problem assignments.)

Why do changes in this field occur so slowly and in such small quantities? One possible explanation is the fact that mathematics textbooks have mainly been written by men and that the male authors tend to take mathematical problems from domains of their own personal experience and interest. Construction, transportation, and traffic occur very frequently, while medicine or ecology are the exception. Shifts of the emphasis towards topics that would appeal to girls may too quickly be brushed aside as "exaggerations" (here I am talking of my own experience as the only woman in a team of authors). Changes in school textbooks should not only rectify the statistical balance between males and females but also reflect more generally the question of what types of mathematical problem-solving are relevant for girls — and for boys (Niederdrenk-Felgner, 1995).

Some alternative topics and problem-solving assignments have been proposed by Jutta Öhler for mathematics instruction in the lower grades of secondary schools. Her collection not only contains stimulating suggestions for lessons but makes it very clear to what extent many areas of everyday life — for example, nursing, children's education, ecology in the home, ecology of transport and traffic, roles in the family, gender roles in jobs — are excluded from "normal" textbooks (Öhler, 1991). Other concrete proposals for different types of problems have been published by an association called MUED (*Mathematik-Unterrichts-Einheiten-Datei e.V.*, Mathematical Education Units Database Association; the German adjective *mued* [müd] also means "tired"). This is an initiative of mathematics teachers for in-service training and the development, exchange, and experimental use of classroom material outside established channels.

So far, there exists no published study on the influence of mathematics textbooks on gender-role assumptions in the minds of girls and boys. A comparative study — where one class is taught using a traditional textbook and another using a textbook that was written fully in conformity with equal opportunities standards — would not be possible at the moment, as textbooks of the second category do not exist in Germany.

Sybille Schütte confronted girls with different types of assignments, and her published results indicate that the relevance of the problem context and the question of whether the problem makes sense to them does influence the girls' ability to come to terms with the traditional textbook assignments (Schütte, 1993).

GENDER AND MATHEMATICS EDUCATION
IN (WEST) GERMANY

It was only in recent years that mathematical educators in Germany began to discuss *gender and mathematics education* on a broader basis. This is surprising, particularly in terms of international comparison, since this has for decades been an established field of research in the English-speaking world. There has also been a long-standing tradition within the educational and social sciences in West Germany of discussing *girls at school*, but this has only very slowly been noted and slipped into discussions on the teaching of specific school subjects.

I would like to outline very briefly the main issues of this educational debate, as they are the foundation of the present debate on gender and mathematics education.

A major stimulus for this discussion came at the end of the '70s from the women's liberation movement. Co-education in the schools, which had only just been introduced in all the West German states, was put into question. Their claim was that experience had shown after the first years that the existing system of co-education — which women had fought for and which had been celebrated as progress — did, in fact, *not* provide equal opportunities for boys and girls in the way it was supposed to do. Co-ed schools were found to be still "boy-oriented," geared to the boys' needs, while no attention was paid to the educational needs of girls. Schools did not deal with different life prospects and life concepts and excluded both the division of labor between the sexes and violence of men against women. This meant that schools did not provide girls with a forum for the discussion of topics that were relevant for them.

This criticism of schools from a feminist point of view has found expression at various congresses on "Women and Schools," which began in 1982. They were initiated by an association set up in 1981 by Ilse Brehmer and Uta Enders-Dragässer, whose aim was to take up the international discussion on sexism and apply it to the German educational system (Brehmer & Enders-Dragässer, 1984; Enders-Dragässer & Fuchs, 1990). From the beginning, this criticism had a broad range of targets including the content of instruction, educational and instructional objectives, schoolbooks, the use of language, and school interaction and hierarchies. This discussion was a forum for an analysis of the existing situation, an exchange of experiences and for alternative concepts. It was only very gradually, however, that these trends reached the reality of the schools.

An interesting turn, in the attitudes of official school administrations and ministries, was brought about by the introduction of computers into the schools. Gender-specific differences in the approach to and the use of computers, different extents of prior knowledge of computers, and the different interests of boys and girls were so evident that an urgency for some sort of intervention was felt. A number of action-research projects were conducted in various West German states (summarized, for example, in Bundesminister für Bildung und Wissenschaft, 1992). One of their motivations was to provide girls with an opportunity to participate right from the start in the shaping of the further development of an important field of modern technology, rather than to remain in the traditional role of mere computer operators, excluded from a whole branch of vocations and careers. And there is, of course, an interest on the part of society, industry, and commerce to train young people of both sexes as early as possible in the use of new technologies.

This discussion of "girls and computers" has led to a revival of the "girls and mathematics" discussion as well. The underlying problems seem to be similar but even more explicit in the case of computers and technology than in the case of mathematics teaching. In some of the German states, lessons dealing with introductory information-technology education have been allocated to mathematics lessons, and in some respect they are based on abilities which mathematics education tries to pass on to pupils.

Various action-research projects and programs on "Girls/Women and Computers" have led to a variety of results that confirm the underlying assumptions of feminist criticism and should lead to practical consequences which are (partly) applicable to mathematics education as well (cf. Niederdrenk-Felgner, 1993, p. 99ff.):

- educational methods and the forms of classroom work should take into account the needs of girls and the structures preferred by them; group work and a co-operative style of work should be encouraged and actively supported by the teacher

- teachers and students should be made aware of classroom interactions; a suitable tool for this purpose is monitoring guidelines and sheets that may be filled in by the pupils

- special consideration should be given to the content of instruction; it should be oriented to the world and environment in which the boys and girls live and include their personal experiences; opportunities for personal identification with positive models should be made available to girls and boys to the same extent; the curriculum should contain content which allows a discussion of gender relations in our society

- the presentation of people of different sexes in the educational material should conform to the standards of equal opportunities

- the pupils' creativity should be addressed wherever possible

These demands are not new in mathematical education; they were already formulated by Christine Keitel in 1989 (Keitel, 1990).

Gabriele Kaiser-Messmer has compiled a catalogue of areas where research and additional work is urgently needed (Kaiser-Messmer, 1989). One of her points is the need for a solid assessment of the situation of women and girls in mathematics education, with data which would provide a broad base for a further discussion of *gender and mathematics education* in Germany, more in line with the relevance of this topic. The small scale of previous discussions is indeed deplorable. Up to 1989, there were hardly any contributions at annual conferences on mathematical education.

In spring 1989, the Association for Mathematical Education (GDM) established a section Women and Mathematics Education. The section's first meeting was held on 10 November 1989 in Berlin (one day after the fall of the Berlin Wall). Since then, two meetings have been held per year, and the great response right from the beginning has shown that there is indeed demand for such a forum of discussion between colleagues of both sexes from the academic world and from schools.

PROSPECTS

To some extent it has been through the activities of the GDM's Women and Mathematics Education section that the gender and mathematics discussion in Germany has intensified and become more lively. The section gave presentations at conferences on mathematics education and initiated two detailed publications of the *Zentralblatt für Didaktik der Mathematik* (ZDM) early in 1994, outlining the present state of international discussion. But gender and mathematics education is still far from being an established field of mathematics education. It is regarded by many as just another fashionable (educational and) political trend, and those involved in this work — mainly women — still have to struggle for recognition. In part this is due to a tradition in Germany to focus mathematics education on questions of method and content rather than on the more general educational context — the influence of school socialization on learning, for example — which is considered to be of only minor importance.

Some encouraging forms of contact and co-operation have been established with the mainstream of mathematics, where there are some first signs of a feminist critique of the subject (Pieper-Seier, 1993). At the moment, it is still a very small circle within which a new style of discussion on mathematics is emerging, which takes into account the "culture" of mathematics, the conditions of mathematical research, existing hierarchies among mathematicians, and their perception of mathematics as a science.

Women mathematicians from academic institutions in East and West Germany have combined to form a strong section within the newly established association European Women in Mathematics (EWM). It was these women who initiated a panel discussion "Women and Mathematics" at the 1991 Annual Conference of the Deutsche Mathematiker-Vereinigung (DMV), dealing mainly with the working conditions of women mathematicians and the question of why the number of women working in Germany in this field is so small. A network of links and exchanges between these activities is emerging, which will provide a more stable base for future discussions.

Mathematical educators now consider one of their most urgent tasks the integration of questions of gender and mathematics education into the curricula of teacher training and in-service training — a topic which has so far not been mentioned in government legislation. Teacher education at universities and teacher training colleges may, however, include the occasional seminar and lecture on women and mathematics, sometimes initiated and organized by the students themselves. A group of (mainly) women students at the University of Tübingen initiated a series of lectures which dealt with the topic from different angles and presented the current state of discussion (Grabosch & Zwölfe, 1992). But this commitment and awareness of teachers and students has little or no effect on the schools. Cuts in educational budgets have led to a situation where hardly any new teachers are employed.

In some German states (Rhineland Palatinate, North Rhine Westphalia, Lower Saxony), the official system of in-service training for teachers includes an option Girls in Science, Mathematics and Technology. Seminars for Mathematics teachers on Girls and Computers are increasingly taking up problems of gender-specific behavior. The DIFF's project *Girls and Computers — Models for More Girl-friendly School Teaching* was accompanied by various seminars where we found that it can be very difficult to open the eyes of (especially) men-only groups and to make them aware of and sensitive to the problem. Lectures on topics like Gender and Giftedness and Effects of Co-education on Girls (and Boys) often met with an emotional and biased response. As a result of these experiences, we have developed a concept for such seminars based mainly on group work and on participants discovering for themselves the essential background needed. This has proved more successful.

Schools play a key role in influencing young people's attitudes to mathematics. Schools are the places where changes are needed to bring about changes in young people's interest in mathematics, in gender-specific attributions, and in the reality of society. It is the schools where (if anywhere) "feminine" roads to mathematics may be opened up by changing approaches and methods, by presenting personalities as examples and, perhaps, by presenting different content. But this will require a more open discussion on gender and mathematics education and the acceptance of the need for such a discussion in Germany.

NOTE

This article has been translated by Lothar Letsche.

REFERENCES

Beerman, L., Heller, K. A., & Menacher, P. (1992). *Mathe: nichts für Mädchen? Begabung und Geschlecht am Beispiel von Mathematik, Naturwissenschaft und*

194   CORNELIA NIEDERDRENK-FELGNER

Technik. Bern: Huber.

Benbow, C. P., & Stanley, J. C. (1980). Sex differences in mathematical ability: Fact or artifact. *Science, 210,* 1262–1264.

Birx, E. (1988). *Mathematik und Begabung. Evaluation eines Förderprogramms für mathematisch besonders befähigte Schüler.* Hamburg: Krämer.

Brehmer, I., Küllchen, H., & Sommer, L. (1989). *Mädchen, Macht (und) Mathe. Dokumente und Berichte 10 der Parlamentarischen Staatssekretärin für die Gleichstellung von Frau und Mann,* Düsseldorf.

Brehmer, I., Enders-Dragässer, U. (1984). *Die Schule lebt — Frauen bewegen die Schule. Dokumentation der 1. Fachtagung in Giessen 1982 und der 2. Fachtagung in Bielefeld 1983 Frauen und Schule. Arbeitsgemeinschaft Elternarbeit (Hrsg.), Band 12,* München: Deutsches Jugendinstitut.

Bundesminsiter für Bildung und Wissenschaft (Hrsg.). (1992). *Mädchen und Computer — Ergebnisse und Modelle zur Mädchenförderung in Comuterkursen. Schriftenreihe Studien zu Bildung und Wissenschaft, Band 100,* Bonn

Dick, A. (1991). *Rollenbilder von Männern und Frauen, Jungen und Mädchen in Schulbüchern. Anregungen zu ihrer Behandlung im Unterricht der Primarstufe und Sekundarstufe I.* Wiesbaden: Hessisches Institut für Bildungsplanung und Schulentwicklung (HIBS), Sonderreihe Heft 30.

Enders-Dragässer, U., Fuchs, C. (1990). *Frauensache Schule. Aus dem deutschen Schulalltag: Erfahrungen, Analysen, Alternativen.* Frankfurt/M: Fischer.

Fennema, El. (1978). Sex-related differences in mathematics achievement: Where and why. In J Jacobs (Ed.), *Perspectives on women and mathematics* (pp. 1–20). Columbus: Ohio State University/ERIC.

Frasch, H., & Wagner, A. C. (1982). "Auf Jungen achtet man einfach mehr..." Eine empirische Untersuchung zu geschlechtsspezifischen Unterschieden im Lehrer/inenverhalten gegenüber Jungen und Mädchen in der Grundschule. In I. Brehmer (Hrsg.), *Sexismus in der Schule. Der heimliche Lehrplan der Frauendiskriminierung* (pp. 260–278). Weinheim: Beltz.

Glötzner, J. (1991). Michael ist der Allergrösste — Rollenfixierungen in neuen Mathematikbüchern. In C. Niederdrenk-Felgner (Hrsg.), *Mädchen im Mathematikunterricht. Arbeitsbericht Mathematik/Informatik* (pp. 35–46). Tübingen: DIFF.

Grabosch, A., & Zwölfe, A. (1992). *Frauen und Mathematik. Die allmähliche Rückeroberung der Normalität?* Tübingen: Attempto.

Hannover, B. (1992). Spontanes Selbstkonzept und Pubertät. Zur Interessenentwicklung von Mädchen koedukativer und geschlechtshomogener Schulklassen. *Bildung und Erziehung, 45,* 31–46.

Heller, K. A. (1990). Zielsetzung, Methode und Ergebnisse der Münchern Längsschnittstudie zur Hochbegabung. *Psychologie in Erziehung und Unterricht, 37,* 85–100.

Hyde, J. S., Fennema, E., & Lamon, S.J. (1990). Gender differences in mathematics performance: A meta-analysis. *Psychological Bulletin, 107*(2), 139–155.

Jungwirth, H. (1990). *Mädchen und Buben im Mathematikunterricht. Eine Studie über geschlechtsspezifische Modifikationen der Interaktionsstrukturen.* Österreichisches Bundesministerium für Unterricht, Kunst und Sport (Hrsg.), Reihe Frauenforschung Band 1, Wien.

Kaiser-Messmer, G. (1989). Frau und Mathematik — ein verdrängtes Thema der Mathematikdidaktik. *ZDM, 2,* 56–66.

Kaiser-Messmer, G. (1993). Results of an empirical study into gender differences in attitudes towards mathematics. *Educational Studies in Mathematics*, *25*, 209–233.

Kaiser-Messmer, G. (1994). Analyse ausgewählter Schulbücher unter geschlechtsspezifischen Aspekten. *Beiträge zum Mathematikunterricht*, pp. 171–174.

Keitel, C. (1990). Mädchen und Mathematik. *mathematica didactica*, *13*(3/4), 31–47.

Linn, M. C., Hyde, J. S. (1989). Gender, mathematics and science. *Educational Researcher*, *18*(8), 17–19, 22–27.

Lopatecki, C., & Lüking, I. (1989). *Bescheiden, sittsam und rein? Rollenklischees in Mathematik-Schulbüchern für die Sekundarstufe I*. Bremische Zentralstelle für die Verwirklichung der Gleichberechtigung der Frau (Hrsg.). Bremen: Selbstverlag.

Niederdrenk-Felgner, C. (1993). Computer im koedukativen Unterricht. Studienbrief Mädchen und Computer. Tübingen: DIFF.

Niederdrenk-Felgner, C. (1994). Was ist schon dabei, wenn Frau Abel den Kühlschrank bezahlt? — Aus der Arbeit an einem Mathematikbuch. *Beiträge zum Mathematikunterricht*, pp. 267–270.

Niederdrenk-Felgner, C. (1995). Textaufgaben für Mädchen — Textaufgaben für Jungen? *Mathematiklernen*, *68*, pp. 54–59.

Öhler, J. (1991). *Mädchen und Mathematikunterricht*. Anregungen und Materialvorschläge zu einem mädchenfreundlichen Mathematikunterricht. Arbeitsberichte aus der Forschungsstelle für Frauenfragen an der Pädagogischen Hochschule Flensburg.

Pieper-Seier, I. (1993). Frauen und Mathematik — ein Forschungsthema? In C. Funken, & B. Schinzel (Hrsg.), *Frauen in Mathematik und Informatik*. Wadern: IBFI.

Schwank, I. (1992). Untersuchungen algorithmischer Denkprozesse von Mädchen. In A. Grabosch & A. Zwölfer (Hrsg.). *Frauen und Mathematik. Die allmähliche Rückeroberung der Normalität?* (pp. 68–90). Tübingen: Attempto.

Schwank, I. (1994). Zur Analyse kognitiver Mechanismen mathematischer Begriffsbildung unter geschlechtsspezifischem Aspekt. *Zentralblatt für Didaktik der Mathematik*, *26*(2), 31–40.

Schütte, S. (1993). Muss ich jetzt plus, minus, mal oder geteilt rechnen? — Von einigen Schwierigkeiten, die Mädchen mit Textaufgaben haben und was wir daraus lernen können. *Mathematische Unterrichtspraxis*, *II, Quartal*, pp. 17–28.

Srocke, B. (1989). *Mädchen und Mathematik*. Wiesbaden: Deutscher Universitäts Verlag.

Turkle, S. (1984). *Die Wunschmaschine. Vom Entstehen der Computerkultur*. Hamburg: Rowohlt.

JOSETTE ADDA

IS GENDER A RELEVANT VARIABLE
FOR MATHEMATICS EDUCATION?
THE FRENCH CASE

The best way to remember the popular ancient French consideration of the women-science relation is to read *Les femmes savantes* and *Les précieuses ridicules* by Molière; this is, unfortunately, more representative than to evoke great women like Emilie de Breteuil, Marquise du Chatelet, or even Sophie Germain (see Dubreil-Jacotin, 1948), and we know that although Marie Curie received two Nobel Prizes, she was never a member of the Academy of Sciences!

Up to the end of the 19th century, girls received separate and very different education than boys; the highest level for girls was elementary teacher training, and even then the textbooks called "Mathematics for Girls" were very shy and simple and used familiar "female contexts" such as embroidery. The point at which change occurred was the Law of Camille Sée (21 December, 1880) which created secondary teaching for girls (Mayeur, 1977); it was only two years ago that Lyon celebrated the centenary of the first baccalaureate obtained by a girl (Julie Daubié). On the 23rd of July 1881, C. Sée succeeded, after long debate, in obtaining parliamentary assent for the creation of the École Normale de Jeunes Filles de Sèvres to train secondary school teachers. It was always considered to be inferior to the École Normale Supèrieure for boys (rue d'Ulm), and so some French women mathematicians such as M. L. Dubreil-Jacotin and Jacqueline Ferrand tried to and succeeded in entering the male École Normale Supèrieure.

In 1924, Leon Bérard's reform identified the secondary curriculum of boys and girls, but schools were left separated. Co-education, or mixed education, in France (Mosconi, 1989) was created only after the World War II (in 1956), first in some *lycées*, and by the end of the sixties in all the French *lycées*. But the recruitment competitions for teachers (*agregations* and *CAPES*) were left separate until many years later and the École Normale Supèrieure became *mixte* with undifferentiated recruitment only in 1986. Unfortunately this revealed a severe distortion in mathematics enrolment, the women numbering now very few. How could it be explained? Some people think it is a "lack of competency," while others explain it by a "lack of orientation." The following analysis will give some answers.

Even when a general equality in education was admitted, there remained a reticence about mathematics. F. Mayeur quotes the following remark by one of the group which wrote the projects for the girls' curriculum in 1880:

G. Hanna (ed.), Towards Gender Equity in Mathematics Education, 197–203.

Les jeunes filles qui, sauf de très rares exceptions, n'iront jamais bien loin dans les études de ce genre, n'auraient que faire de cette logique inflexible qui y est necessaire, et qui serait plutôt nuisible dans la pratique de la vie, où l'on ne parle presque jamais d'idées abstraites et de principes absolus. (Mayeur, 1977, p. 94)

As other girls of my generation, I attended a *lycée de jeunes filles* for secondary schooling and when I intended to do *Mathématiques Supérieures* in Nice, I had to go to the *Lycée de Garçons*, where I was one of three girls in a class of 40 students. Next year I went to Paris in a class of *Mathématiques Spéciales* for girls in the Lycée Fenelon. For all my studies in these feminine schools, I had women teachers. In university I had the honor of working with M. L. Dubreil-Jacotin. During her algebra lessons, she referred very often to E. Noether. So I never felt any problem in my relationship to mathematics as a woman. I only considered that the job of engineer, as a job in a factory, was a "male job" and that this was the reason why so few girls were engaged in mathematics studies. For me, there was no problem pursuing mathematics teaching or research! Through my involvement in ICMI conferences, and especially in the IOWME Group, I have come to understand that things are not so simple. I often hear that for studies in mathematics education, some people are taking into account the gender variable. In this chapter, I shall analyse if and how this variable is relevant to the present French situation.

The official published French statistics present two aspects of the relationship between mathematics education and gender. The first aspect is that of performances, and the second aspect is orientations and choices. As we shall see, it is plain that gender is not a relevant variable to the study of the first aspect, whereas it is relevant to the second aspect. So it is important to interpret and explain the imbalance between these two facts.

IS GENDER A RELEVANT VARIABLE FOR PERFORMANCES?

National assessments have been recently organized by the Education Ministry to give a better understanding of the French school system. They take place at the beginning of each school year (September) as students enter the CE2 (third class of primary school, with children normally eight years old), the sixth class (the first year of the secondary school, with children normally 11 years old) and the second class (15 years old). Unfortunately, for political reasons, the results of the assessment for this last level are not known at the national level. In 1990, an inquiry using a representative sample was also done of students entering the third class (students normally 14 years old). Table 1 (from DEP, 1991) lists some of the national results of this year (Adda, 1993; Adda, Goldstein, & Schneps, 1994; Baudelot & Establet, 1991; namely a synthesis of the average difference between the scores of girls and boys (for total scores of 100) in French and in mathematics:

Table 1: Écarts moyens des scores obtenus en français et en mathématiques
par les filles et les garçon toutes catégories sociales confondues

	Français	Mathématiques
CE2	5,6	-0,2
Sixième	6,0	0,5
Troisième	4,5	-4,3

Les écarts sont rapportés à des scores totaux tous également remenés à 100.

So, it appears that at these three levels, the difference between scores of boys and girls in mathematics is very small (it is positive or negative depending of the fields of items), whereas gender appears very influential in French, with girls doing much better than boys, but who is worried about "Gender and Language Education"?

The Department of Education also did some analysis taking into account the socio-economic status of the families of students; the synthesis (see

Table 2: Un préjugé tenace ne résiste pas à la rigueur
de l'analyse de la variance

Pourcentage de variance expliquée (R^2) par le sexe, l'age et la catégorie sociale du père de l'élève

	Français	Mathématiques
CE2	18,7	13,1
Sixième	28,0	19,5
Troisième	9,1	18,6

Part relative du sexe, de l'age et de la catégorie sociale du père de l'élève dans le pourcentage de variance expliquée.

		Français	Mathématiques
Le sexe	au CE2	13,7	5,7
	en Sixième	11,3	0,2
	en Troisième	35,1	3,6
L'age	au CE2	45,7	46,4
	en Sixième	67,3	78,4
	en Troisième	50,2	73,5
L'origine sociale	au CE2	40,6	53,6
	en Sixième	21,4	21,4
	en Troisième	14,7	22,9

details and comments in publications listed above) appears in Table 2 on the previous page.

These figures show that for mathematics performance (but not for French performance), gender is not at all a relevant variable, at least at these school levels; the relevant variables are age and socio-economic status (and many studies have shown the links between those two variables).

A question may be asked: Is it possible that this equality of the performance of students up to the age of 14 could vanish at an older age? As I said earlier, no national result is available for the assessment in the second class (15-year olds), but by considering the results of some schools (perhaps not representative), I've arrived at the hypothesis that there is again, at that level, no significant difference in performance between girls and boys in mathematics. In her DEA dissertation, F. Agricole (1994) analysed the answers of approximately 300 students. We cannot consider that analysis to be scientific, because there is no guarantee that the schools considered are representative, but Agricole can conjecture that the situation is similar to the lower levels. There appears to be no clear superiority of one gender over the other in her sample, as some questions were better answered by girls, some by boys, which sometimes happened in one classroom while the inverse occurred in another classroom.

IS GENDER A RELEVANT VARIABLE FOR ORIENTATIONS AND CHOICES?

There is a very shocking contrast between the figures quoted above showing the abilities of girls in mathematics before the levels of orientation and the figures below showing the reality of the students' choices after orientation.

Table 3 gives the distribution of students according to gender in the sections of the final class of secondary school (preparing the "baccalaureate") in 1990/91 (DEP, 1992). The sections with the most mathematics are C and E. Section D is scientific but emphasizes biology; A and B are the literary sections; F, G, and H are the technical ones.

It is interesting to represent (Adda, 1993) these numbers graphically in two ways: the distribution of 100 students of each gender by section (Figure 1) and the distribution of 100 students of each section by gender (Figure 2).

INTERPRETATION

In our modern world, it is clear that these orientations are not equivalent; the better job possibilities are linked to the scientific orientations. So the scales of the last graphic are very similar to the social scale, and that "choice" seems very difficult to understand considering the equality of possibilities. How can we explain that distortion? I think that it is the conse-

Table 3: Le second cycle général et technologique par section

	Public			Privé			
	Garçon	Filles	Total	Garçon	Filles	Total	Total
Terminale:							
A	12 998	58 392	71 390	4 567	15 299	19 866	91 256
B	27 087	44 218	71 305	10 127	12 047	22 174	93 479
C	37 517	21 688	59 205	8 701	4 515	13 216	72 421
D	29 775	30 917	60 692	10 695	8 748	19 443	80 135
E	9 970	639	10 609	674	57	731	11 340
F	36 228	13 732	49 960	4 888	5 478	10 366	60 326
G	27 103	55 101	82 204	7 820	12 666	20 486	102 690
H	139	65	204	196	50	246	450
Brevet de technicien	7 712	2 181	9 893	979	940	1 919	11 812
Total	188 529	226 933	415 462	48 647	59 800	108 447	523 909

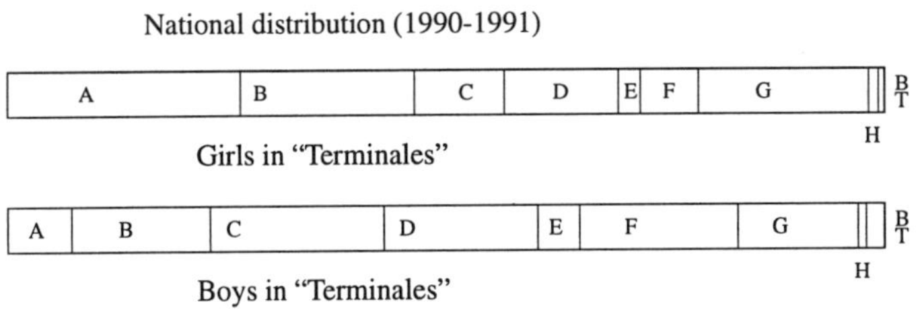

National distribution (1990-1991)
A
B
C
D
E
F
G
H
B
T
Girls in "Terminales"
A
B
C
D
E
F
G
H
B
T
Boys in "Terminales"

Figure 1

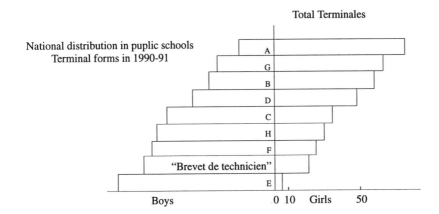

Total Terminales
National distribution in puplic schools
Terminal forms in 1990-91
A
G
B
D
C
H
F
"Brevet de technicien"
E
Boys
0
10
Girls
50

Figure 2

quence of the conjunction of various prejudices:

Prejudices of the parents: French parents often feel a good job is less necessary for girls than for boys (they think that girls will be taken care of by their husbands) and when parents have to make financial choices (for example, to pay for additional mathematics lessons), they select brothers over sisters.

Prejudices of the teachers: During the school years, orientations are the result of decisions made by teachers' councils, and it appears that teachers (both male and female) require more from girls than from boys for the scientific orientation. The best proof of this is that the proportion of success in baccalaureate C is higher for girls than for boys. Gasquet and Ruffieux (1990) note that, in some places, the girls of section C are so over-selected that they succeed 100 percent!

Prejudices of the employers: The inequality of judgment of teachers is not completely absurd; it is true that, when applying for a job, a female has to be better than a male to obtain the job, and if they have equal qualifications, it is well known that the employer will choose the male. There is always the argument about absences for maternity and, now, in some places in France, it happens that to obtain a job, women have to swear that they will not become pregnant in the first two or three years. This is, of course, completely illegal!

Prejudices of students: All the prejudices above are integrated by the students and combined to push girls to renounce scientific studies. There also exists some specific prejudices well known in the community of mathematics educators — especially the image of femininity, the "fear of success," and the other psycho-sociological bridles studied by E. Fennema and G. Leder in many of their publications. In fact, all those prejudices contribute to the constitution of gender as a "social condition," as E. Fennema said in her lecture at the ICMI study meeting.

   Is it possible to change the situation? All prejudices can be destroyed by information and knowledge. The book *Allez les filles*, written by the sociologists Baudelot and Establet (1991) has had popular success. The French association Femmes et Mathématiques takes opportunities to present the problem in newspapers and particularly in women's magazines. This association, in conjunction with the Association of Mathematics Teachers (APMEP), whose president is often a woman, organizes information meetings directed to mathematics teachers and secondary students (Adda, 1993; Adda, Goldstein, & Schneps, 1994). The French government has also decided to become involved by launching national information campaigns; in schools there are posters and documents (papers, booklets, and even a video) especially in guidance offices, telling girls *"c'est technique, c'est pour elle*" ("it is technical, it is for her") and showing girls happily doing technical and scientific jobs. A national scholarship was created in 1986 to help girls going into scientific or technical studies. For the past three years, it has been changed to a "Prix de la Vocation" awarded to 40 girls in each region, selected from students in the final year of secondary school who want to continue their studies in the scientific or technical domain. As a member of the jury of that prize, I can say that it is very interesting to read the essays of these girls. It is clear that they are aware that to succeed in their science studies is not enough for a girl; they are very conscious that they will also have to fight against many social constraints and prejudices.

REFERENCES

Adda, J. (1993). Les filles et l'enseignement des mathématiques. La situation en France et à l'étranger. In *Actes du Colloque La femme dans la société: choix et contraintes de sa vocation* (Laboratoire d'ethno-éducation comparée). Université Rennes 2 – Haute-Bretagne.

Adda, J., Goldstein, C., & Schneps, L. (1994). France. In L. Burton, (Ed.), *Who counts? Assessing mathematics in Europe: European Colloquium on differential performance in the assessment of mathematics*. Stoke-on-Trent: Trentham Books.

Agricole, F. (1994). *Etude sur les performances des élèves lors de l'évaluation à l'entrée en seconde. Mémoire de DEA de Didactique des Mathématiques*. Université Paris 7.

Baudelot, C., & Establet, R. (1991). *Allez les filles!* Paris: Editions du Seuil.

Direction de l'Evaluation et de la Prospective (DEP). (1991, Aout). *Revue Education et Formations*, 27–28.

Direction de l'Evaluation et de la Prospective (DEP). (1992). *Repères et références statistiques sur les enseignements et la formation*. Ministère de l'éducation nationale et de la culture.

Dubreil-Jacotin, M.L. (1948). Figures de mathématiciennes. In *Les grands courants de la pensée mathématique*. Paris: Editions des Cahiers du Sud.

Gasquet, S., & Ruffieux, M. (1990). *Lycée: Peut mieux faire!* Paris: Ed. Syros Alternatives.

Mayeur, F. (1977). *L'enseignement des Jeunes Filles sous la Troisième République*. Paris: Presses de la Fondation nationale des sciences politiques.

Mosconi, N. (1989). *La mixité dans l'enseignement secondaire: un faux-semblant?* Paris: Presses Universitaires de France.

Mosconi, N. (1994). *Femmes et Savoir, La société, l'école et la division sexuelle des savoirs*. Paris: L'Harmattan.

ROSA MARIA SPITALERI

WOMEN'S KNOW–HOW AND AUTHORITY: ITALIAN WOMEN AND MATHEMATICS

The relationship between women and mathematics in Italy seems to be especially lucky. While female participation in mathematics education and mathematics-based careers in other countries is still very limited, in Italy the proportion of women attending university exceeds men, and information on mathematics-related careers shows a consistent number of women achieving professional advancement (see Barucci, 1993).

However, we would like to ask what these advances in the mathematics field mean. Have all the barriers that account for women's under-representation and gender-based inequities been removed? It seems there are no physical nor intellectual barriers to women's achievement since women's capability in mathematics is a concrete thing: girls and boys demonstrate equal ability in teaching and learning mathematics. Nor do social and cultural barriers seem to affect aptitude and aspirations since women are successful in high-level mathematical courses and professional careers.

However, several problems remain which relate to women's feelings about working in traditionally male-dominated professions such as mathematics. What specific identities have women had to assume to be accepted in the mathematical community? How do they assess their own aptitude and are they identified, in a psychological sense, with their ability and achievement? Have women had an effect on the culture of scientific life in Italy, and what kinds of relationships are they able to build and promote? These are a few of the open questions we will discuss. That is, we will be analysing the "fit" between Italian women and mathematics and we will be providing a more detailed overview of the encouraging prospects for women in mathematics in Italy.

WOMEN'S KNOW–HOW

It seems important not only to recover quantitative information on women's participation in mathematics but also to focus on the qualitative characteristics of being a woman which may have an impact on women's presence and initiative in the field. Identifying know-how which is specific to women may move women to act with a new consciousness in the mathematical community.

Women's achievement in science appears not to be officially recognized,

G. Hanna (ed.), Towards Gender Equity in Mathematics Education, 205–213.
© 1996 *Kluwer Academic Publishers. Printed in the Netherlands.*

even by women themselves. Living in advanced industrial countries pre-supposes becoming familiar with scientific concepts, methods, and performances as much, if not more, for women as for men. As I have argued (Spitaleri, 1994), women can be easily identified as

- *end-users of scientific theories and results* (old and new technological tools were and are the base of women's freedom from heavy housework; medical science provides support in saving children's lives, etc.)
- *integrators* and often *developers* of scientific knowledge (the medical role given to women by society requires eyes trained to discover diseases, experience in coupling psychological and medical therapeutics, and other fundamental functions)
- *scientific operators* (women have access to and operate in scientific fields playing professional and educational roles)
- *symbolic models* (mothering has been used to model an approach to scientific knowledge which emphasizes taking care of all the subjects involved in the knowing process — see Donini, 1991)

Thus women are very involved in science and they possess scientific expertise at several levels. Does this relationship between women and science extend to mathematics and to mathematics-related fields, such as computer and technological studies?

Teaching mathematics has been traditionally charged to women in Italy, and probably every Italian student has had at least one female teacher of mathematics. Indeed teaching seems to be a female task for all the disciplines, especially at the secondary school level. From women's point of view, this widespread practice could become a strength. Women could enlarge their task to assume a fundamental role in young people's education and have an effect on both their culture and their future aspirations. Teaching is one of the professional areas in which women could challenge society and recover the authority of educational expertise. The presence of women in education could be determinant in increasing male and female interest in mathematics. However, teaching does not presently have social prestige, for a complex set of reasons, and women are forced to accept a role bereft of authority.

In daily life, women show a special aptitude for the "optimization" of tasks. From youth, they need to take into account different, parallel requirements, provide simplified procedures to interact with complex realities and events, and optimize timing of task execution. I would like to recall the sentence "Parallel processing seems to be thing for women" (see Hay, 1993). This essential female capability could be useful in work environments where separate scientific approaches to complex problems have to be evaluated, compared, and eventually composed in new procedures.

Capabilities developed mainly by women — educational expertise, competence in mathematical studies, and experience in problem solving in complex environments — should be recognized in women's careers. Decision-making positions, still reserved for men, should be open to responsible female management. On the other hand, studies of women's know-how should be promoted. By identifying what are hidden but well-used women's capabilities, roles charged to women can regain prestige. The scientific thought and space they now occupy in isolation could provide a new framework for understanding the dynamics of society and, in particular, educational and research projects.

WOMEN'S PARTICIPATION

A few years ago, I was looking for data on girls and boys attending universities. The consistency and degree of female participation in mathematical studies was a pleasant surprise: in fact more women participate in mathematics than men. The data in Tables 1 and 2 track male (M) and female (F) students at the University of Bologna in the academic years 1990/91 and 1991/92. The three columns of data contain, respectively, the total number of students studying mathematics, the number starting their studies, and the number passing all their examinations by the required time. The data show not only a greater number of females in mathematics but also about equal or even better mathematics achievement, if regular course attendance (90% to 89% for females compared to 80% to 83% for males) can be a measure of aptitude. The data encouraged me to make the conjecture that positive female attitudes towards mathematics really exist and to investigate national trends of analogous data.

Table 1: Students in mathematics at the University of Bologna

UNIBO 90/91	Total students	1st year students	Total regular students
Females	481	165	431
Males	275	83	221

Table 2: Students in mathematics at the University of Bologna

UNIBO 91/92	Total students	1st year students	Total regular students
Females	534	135	475
Males	288	74	239

Analogous national data related to the academic year 1992/93 are presented in Table 3. They allow my conjecture to gain consistency, even if the percentage of regular students decreases to 71 percent for females and 70 percent for males, making both groups about equal. On the basis of this data, I felt it would be useful to follow the complete trend of data related to university students in order to support hypotheses about women's ability in mathematics.

In the same academic year, 1992/93, females comprised 66 percent of the students attending the 30 Italian universities that teach mathematics. For the cohort of first-year students, females comprised 62 percent. Of the 1391 students who graduated in mathematics, 985 were women (71%).

Table 3: Students in mathematics at Italian universities

UNI 92/93	Total students	1st year students	Total regular students
Females	11.685	2.846	8.283
Males	6.099	1.757	4.249

The data clearly show that more women than men are studying mathematics at universities. Moreover, the percentage of women in professional careers remains consistently high. Data related to different research institutions appear in the following tables. The number of women and men belonging to the three levels of professional life in Italian universities appear in Table 4 (updated September 1994).

Table 4: Teaching staff in mathematics at Italian universities

UNI	Researchers	Associated professors	Professors
Females	319	352	82
Males	355	573	569

These data have been recovered from the databank service IONIO of CINECA (see Cocchi, 1994). This special databank, a project of the University Educational Program, imports data from the Istituto Nazionale di Statistica which monitors Italian universities. The bank can be consulted online by network and contains quantitative information on students, graduates, and teaching staff.

Women also seem to be pursuing careers in public research institutions. I would like to give special attention to the data collected about women and men in research positions at the Consiglio Nazionale delle Ricerche (CNR). The data in Table 5 (updated in 1992) are related to all fields of CNR research.

Table 5: Research Staff at the CNR

CNR	Researchers	Senior researchers	Research directors
Females	451	269	49
Males	863	757	275

Table 6: Research staff in mathematics at CNR

CNR	Researchers	Senior researchers	Research directors
Females	12	15	2
Males	16	6	7

Researchers working in mathematics represent about 2 percent of CNR's global staff; in seven institutes, there are 59 males and 44 females employees and of the 58 researchers half are women. Table 6 shows female involvement in the Mathematics Committee of the CNR for each of the three levels of researcher (updated July 1993).

It would be interesting to have information on other aspects of women's investment and production in mathematics over the same time period. For instance, it would be interesting to know what percentage of the 568 publications written by Italian researchers in well-known international journals in 1990 were authored by women. These publications represent 3.6 percent of the mathematical production of the world. Further, Italian mathematicians, in comparison to Italian researchers in all other fields, achieved the best publication results (see ISRDS-CNR, 1993).

WOMEN AND EDUCATIONAL INSTITUTIONS

With respect to the relationship between girls and mathematics, the positive picture I have presented of the Italian case, based on enrolment and career data, begins to change when official competitions are considered. A very central problem here is how to evaluate capabilities at all levels of learning. Moreover, analogous problems come up in the evaluation of scientific production for professional advancement.

Data coming from school competitions — results of the national and international selections of the best students, called "Olympiads" — offer a different view. Girls are so rarely winners that a special prize is reserved for the best girl mathematician. In the 1991 Olympiad, the final Italian selections consisted of 33 females and 293 males; in 1993 there were 48 females and 300 males. Searching the results of recent years for a female winner, I found just a few girls among the best ten students and no girls among the best six who went on to participate in the international competitions.

This information coupled with what we know about the "good" scholastic behavior of girls could confirm a common perception — that is, women can be diligent but not brilliant students, at least in the way men can be.

I believe it is very important to analyse evaluation criteria characterizing school behaviors and competitions and investigate what criteria assume lower scores for women. For instance, answering questions without hesitation can influence one's performance under the stress of competition and reward this habit of boys and men. Thus, redefining the criteria for evaluation and analysing their location vis-à-vis women appears crucial if we are going to obtain official prestige for women's educational achievements and update the criteria to incorporate women's values.

It appears that social and cultural barriers, which may have seemed to disappear, can still jump out in renewed forms.

The Italian institutions involved in planning education are the Ministry of Education which determines secondary school programs, the Ministry of University and Scientific Research, along with the Consiglio Universitario Nazionale (CUN), which determines, with the support of other institutions, university courses, doctoral programs, and definitions of what constitutes mastery in a discipline.

The CNR, for instance, has been providing scholarships, mainly in mathematics, for both undergraduate and graduate students working on original theses. This institution has a direct relationship with the university community since most members of the advisory committees of CNR, which make recommendations on planning and funding Italian research, are professors, associated professors, and researchers working at the university. University and CNR representatives on these committees are designated by separate elections.

The investigation of women's positions in professional careers and decision-making organizations in the mathematics field shows a very strange global picture in light of the large numbers of women studying and graduating in mathematics (see ISTAT data in Cocchi, 1994). Regarding the teaching of mathematics in secondary schools, I stress that the authority of teachers, who are mostly women, needs to be regained. Moreover, women should be encouraged to achieve positions in the management of educational planning and organization.

Looking at the current staff of both the CNR and the 30 Italian universities (see Tables 3, 4, and 5), we notice that the number of women in decision-making positions appears to decrease as the status of the position increases. There are a very small number of women, or none at all, in the highest positions. The decision-making committee of the CNR, for example, consists of nine members who provide leadership for seven CNR institutes and for four national groups of researchers in mathematical areas. The committee composition, as the result of the last election, contains just one woman, a researcher from the university. Moreover, all the heads of the seven institutes of the committee are men, six from the University.

It would be interesting to recover a large set of data quantifying current professional positions of women in mathematics, develop analyses of women's relationship to power, and identify the main factors affecting their feeling towards scientific authority (see Ipazia, 1992; Tedeschini Lalli, 1992).

WOMEN'S ASSOCIATIONS

Women should be encouraged to contribute all their experience, both formal and informal, to scientific practice and thought. Women's associations can provide support for women's reflections and analyses, create national and international communication networks, and allow women to get out of their traditional isolation (see Barucci, 1993; Spitaleri, 1994).

Indeed, existing organizations in mathematics and mathematics-related fields appear to be male controlled. Several women's associations in separate professional fields have recently been created — for instance, astronomy, mathematics, and economics.

European Women in Mathematics (EWM) and Associazione Italiana Donne in Matematica (AIDIM) were created to look at new initiatives for improving the identity of girls in mathematics and for removing gender inequities from the mathematical community.

The EWM was born during a panel discussion organized by the American Association for Women in Mathematics (AWM) at the International Congress of Mathematicians, held in Berkeley in 1986. The aim of this organization has since been to provide a forum for professional and personal analyses and to promote mathematical studies among women. Six EWM congresses have been held, in Paris (1986), Copenhagen (1987), Warwick (1988), Lisbon (1990), Marseilles-Luminy (1991), and Warsaw (1993). The next will be held in Madrid (1995). Members of EWM come from the following countries: Austria, Belgium, Bulgaria, Czechoslovakia, Denmark, England, Finland, France, Germany, Greece, Holland, Hungary, Italy, Lithuania, Norway, Poland, Portugal, Russia, Spain, Sweden, Switzerland, and the Ukraine. Every country can independently organize its appropriate national association and interact with the European Association via the EWM national co-ordinator.

The AIDIM was born in Rome on May 11, 1993 with the following aims:

- to promote women's study and research in mathematics
- to support women in starting or continuing in careers in mathematics
- to promote studies of women's role in the knowledge, transfer, and application of mathematics
- to improve opportunities for the presence of women in the mathematical community
- to co-operate with organizations having analogous purposes in Europe and elsewhere

During a plenary session of the IMACS Congress 1994, held in Atlanta, I understood a basic reason for women's associations, and new aspects of this thought were developed at the AWM Workshop, held in San Diego the day before the 1994 annual conference of the Society for Industrial and

Applied Mathematics (SIAM). The small number of women participating in the IMACS Congress was a concrete and visible thing, reproducing the atmosphere of many other analogous congresses. Moreover all the meaningful scientific relationships were completely managed by men.

The AWM workshop was the other face of the coin — that is, just a few men participated, all the speakers were women, awards were assigned to the best five students from different American universities, and all the scientific relationships were managed by women. Both situations reinforced the idea that women's associations are a very useful tool for women trying increase women's scientific authority and their own scientific relationships, and built more balanced working communities.

WOMEN'S AUTHORITY

As I suggested earlier in the chapter, scientific method is not so far removed from women's approach to the solution of complex problems in both work and daily life. Women's methods of operating are not so different from those of experimental studies. Therefore it seems that their main difficulty is having a chance to build the relationships by which hypotheses and processes provide scientific results.

Indeed women themselves seem to decline positions of authority and help official institutions to discourage girls from choosing careers or pursuing professional levels that require both high scientific expertise and management abilities (see Ipazia, 1992). Yet women should believe in the possibility of having an effect on the scientific community and exerting special control over scientific trends and financial support. They could propose and experiment with new styles of work and change rules and evaluation criteria. Qualitative more than quantitative evaluation of scientific production is needed to allow new concepts and practices to emerge. It appears women are essential to rethinking the life habits of the scientific community.

On the one hand, science appears to be the only authority in modern society. On the other hand, science is a social enterprise which affects and is affected by the relational and interactional structure of current society (see Ipazia, 1992). Do women want to gain authority in this enterprise? Is the price to pay too high? Does having access to scientific neutrality mean forgetting to be a woman? Many questions should be investigated to understand women's feelings and perspectives.

Women also need to review what having authority means. Does having authority begin by recognizing expertise both in ourselves and other women? Does it mean discussing trends of modern science? Does it imply teaching several scientific/mathematics approaches and developing the critical ability of young people to activate their creativity?

In dealing with scientific debate, a fundamental question is the revaluation of different approaches to our thinking. Mathematical abstraction translates into general and formal statements by increasing the distance from concrete objects. However, abstract thinking is not the only way of thinking required by mathematical research in, for instance, applied and industrial mathematics. Visual thinking provides a powerful approach to overcoming difficulties met by traditional mathematical methodologies. Several tools for educational or research purposes are based on the revaluation of visual attitudes of students and scientists (SIGGRAPH, 1993).

Does renewing the meaning of scientific authority mean moving towards a pluralistic approach to study, teach, and develop mathematics (see Turkle & Papert, 1992)? Does it also mean women have to introduce themselves to the scientific committees of well-known international journals?

Women in mathematics will answer these and other questions by their thought and their concrete advances in professional careers and they will transfer their efforts to the next generation.

REFERENCES

Barucci, V. (1993). Donne e matematica. *Lettera Pristem, 9*, 53–54.

Cocchi, A. (1994). Dati sull'istruzione universitaria. *Notizie dal CINECA, 19*, 8–9.

Donini, E. (1991). *Conversazioni con Evelin Fox Keller- una scienziata anomala.* Milano: Eleuthera.

Hay, A. (1993). Do computers separate men from women? Thoughts provoked by Turkle and Papert. *Journal of Mathematical Behavior, 12*, 205–207

Ipazia. (1992). *Autorita' Scientifica Autorita' Femminile.* Roma: Editori Riuniti.

ISRDS-CNR. (1993). Scienza e tecnologia in cifre. Roma: CNR.

SIGGRAPH. (1993). Panel: Visual thinkers in an age of computer visualization problems and possibilities. *Computer Graphics Proceedings.*

Spitaleri, R. M. (1994). Looking for women's know-how in science. *Atti SIMAI94.*

Tedeschini Lalli, L. (1992). Fatti, numeri e femmine. *Sapere, 12*, 45–50

Turkle, S., & Papert, S. (1992). Epistemological pluralism and the reevaluation of the concrete. *Journal of Mathematical Behavior, 11*, 3–33.

TERESA SMART

GENDER AND MATHEMATICS IN ENGLAND AND WALES

The need to push more British girls into mathematics and science reached the top of the national agenda in 1976 when Prime Minister James Callaghan asked: "Why is it that such a high proportion of girls abandon science before leaving school?" Fifteen years later, this question was firmly off the agenda, and a series of initiatives which had helped girls had been withdrawn or reversed. This article focusses on the issue of gender and mathematics in England and Wales. It charts the progress made by girls in mathematics and how and why this progress was halted and turned back in such a short time.

In the 1970s, industry had made a successful public case that it needed more skilled labor. More young people were to be encouraged to move into science and engineering, and it was accepted that this meant drawing in girls who had previously been excluded. At the same time, the developing women's movement was raising awareness that girls and boys did not have the same educational opportunities. Thus, in the late 1970s, there were many initiatives aimed at redressing the gender imbalance in mathematics and science.

These continued under the Conservative government that took power in 1979. In a 1985 government paper *Better Schools*, Education Minister Keith Joseph wrote of schools which at their best "bring out what the pupils have to give by setting challenging goals based on high expectations and by motivating them towards active, well ordered enquiry rather than passive learning" (DES, 1985).

However, in the late 1980s, there was a sharp change; the language and ideas promoted in *Better Schools* disappeared. This was heralded by the closing down of the Inner London Education Authority (one of the most progressive education authorities) and the introduction of the Education Reform Act (ERA). From 1988, there were active moves to take away the gains girls had made in mathematics.

This article focusses on the situation of gender and mathematics in England and Wales. Scotland is the third part of Britain, which with Northern Ireland constitutes the United Kingdom. Scotland and Northern Ireland have somewhat different education systems; these are affected by the debate and changes reported here, but changes were introduced first in England and Wales and are, in general, more radical there. The statistics collection is inconsistent, some data refer to England and Wales, others to Britain, and still others to the UK.

G. Hanna (ed.), Towards Gender Equity in Mathematics Education, 215–236.
© 1996 *Kluwer Academic Publishers. Printed in the Netherlands.*

Equal opportunities in mathematics is not only concerned with issues of gender but also of race and class. There is a significant under-performance of black pupils (both boys and girls) in mathematics relative to white pupils. Research has shown the difference in performance of black pupils is already significant in mathematics in the infant school, particularly for black boys, and in junior schools, where both girls and boys are underachieving (Plewis, 1991). This trend continues to grow throughout secondary school (Drew & Gray, 1990). The Girls and Mathematics Unit at the University of London's Institute of Education found in their research that "the strongest performance differences were in terms of social class, not gender" (ESRC, 1988). Two important books which challenge the male Eurocentric view of mathematics and which are part of the process of reforming the mathematics curriculum to fight for social justice and equality are George Joseph's book *The Crest of the Peacock* (Joseph, 1991) and Sharanjeet Shan and Peter Bailey's book *Multiple Factors* (Shan & Bailey, 1991). However, as I note in several places, the issue is one of biases toward white, middle- and upper-class boys; the high barriers which restrict black and working-class children often affect all girls, so gender should not be seen in isolation.

1980 TO 1988: GENDER ON THE AGENDA

In the mid-1980s as a mathematics teacher working in a school in London, I felt that issues of equality and social justice were foremost in the lives of any progressive teacher. Central government had passed laws against discrimination, notably the Sex Discrimination Act (1975) and two race-relations acts (1968, 1976). The Sex Discrimination Act (SDA) made it unlawful to discriminate against women and girls in education. The SDA said that girls and boys must have equal access to the same curriculum. Responsibility for SDA implementation rested with local education authorities, governing bodies, and heads of schools and educational establishments. The role of administering and monitoring the implementation was given to the newly established Equal Opportunities Commission. This legitimized the work with girls that was being undertaken by individual teachers. "The response to the letter and spirit of the law by teachers, particularly those working in the metropolitan authorities, provided the basis for a teacher movement. This movement has constituted a major challenge to mainstream educational ideas and practice" (Weiner & Arnot, 1987). There were initiatives aimed at understanding and hence challenging the differential experiences of girls and boys, and black and white pupils. These initiatives were promoted by local education authorities, unions, and pressure groups (Arnot, 1985; Arnot & Weiner, 1987; Headlam Wells, 1985; Taylor, 1985). Classroom teachers "played the central role in challenges to the traditional sexual divisions of schooling" (Weiner & Arnot, 1987). Teachers and progressive education authorities challenged a curriculum they believed ignored, or was biased against, the contribution of black and ethnic minority cultures and excluded the experiences of girls and women.

Harder to challenge was the "hidden curriculum" that "does what the official curriculum is presumably not supposed to do. It differentiates on the basis of sex." (Clarricoates, 1987). As Madeleine Arnot noted "The hidden curriculum of schooling (gender dynamics in the classroom, sexual harassment, gendered youth cultures and traditional teacher attitudes) have not proved amenable to reform through persuasion or legislation" (Arnot, 1987). Why and how schools were failing girls became an issue. Gamma (Girls and MatheMatics Association, later Gender and MatheMatics Association) was formed in 1980 to raise awareness of the issue of girls and mathematics. Similar initiatives were promoted in science (Kelly, 1981; Whyte, 1986; Whyte, Deem, Kant, & Cruickshank, 1985).

Cockcroft Report Provoked Far-Reaching Changes

Aside from these broader changes that were taking place, mathematics teachers were also involved in a major evaluation of their subject. A government investigation into the teaching and learning of mathematics by Sir William Cockcroft, *Mathematics Counts* (Cockcroft, 1982), was far reaching and provoked changes in the mathematics classroom. Two aspects of this report were significant. First, it dealt specifically, although only briefly and apparently as an afterthought in an appendix, with the issue of girls and mathematics. The appendix brought into the open current research on girls' performance in mathematics. Second, it contained a by-now famous paragraph, 243, which put forward the expectation that every mathematics lesson should contain more than "chalk and talk" and should include investigational work, practical, and problem-solving tasks, and discussion between pupils.

Data presented to the Cockcroft Committee showed that girls were underrepresented in public examinations. At that time, pupils could be entered for one of two different public exams at 16, the end of compulsory schooling. Those considered the most able (about 30% of the age group) were entered for O-level (Ordinary level, in Scotland called O-grade). Papers were graded A to E, unclassified and fail; only grades A to C were considered a pass suitable for entry into further study or certain jobs such as teaching. A further 56 percent were entered for CSE (Certificate in Secondary Education) examinations. This qualification was graded 1 to 5; a grade 1 counted as equivalent to a grade C at O-level. In 1979 girls made up only 44 percent of the entry in mathematics at O-level; 17.6 percent of the girls gained a pass at grade A to C compared with 24.5 percent of the boys. Twice as many boys (5.5%) as girls (2.6%) gained the highest grade, grade A.

In 1979, in England and Wales, less than 20 percent of the school population at 16 went on to study two or three A-levels (Advanced level courses) which are taken at 18 and are a necessary qualification for entry into higher education. Far fewer of the girls who had gained a pass in O-level continued to study mathematics at A-level; 48.6 percent of girls yet 66.9 percent of boys. Many of the initiatives and actions that took place later were to try to understand and take action on the opting out and underperformance of girls in mathematics.

Cockcroft gave a seal of approval to an investigational approach to the teaching of mathematics. Teachers felt that an approach that favored discussion, collaboration, and investigation would help girls perform better and gain a greater control over their own learning. Although there was no research specifically in mathematics showing this, "those interested in this field report that girls and boys do appear to participate on a much more equal basis when they are in control of their own learning. And girls themselves are very positive about such experiences" (Burton & Townsend, 1985). This was backed by research in science education (ILEA/OU, 1986, pp. 72–73), which showed that where teachers "used pupil-centred enquiry methods" as a teaching style, "this style was most effective in maintaining girls' liking for science."

In the aftermath of Cockcroft, mathematics teachers were involved in discussion, meetings, and a re-evaluation of their teaching method. Government funds were made available to employ advisory teachers — dubbed "Cockcroft Missionaries" — to facilitate this change.

Promoting Change

The atmosphere was of discussion, action, and change and was ripe for challenging the tendency of girls to underperform or opt out of mathematics study. Gamma was formed in 1980 and became a focus for research and data gathering. It ran conferences such as "Be a Sumbody," attended by over 200 girls, which aimed to "convey a girls-friendly message about the nature of mathematical activity, and to reinforce the narrowing effect on career prospects of disengaging from school mathematics" (Burton & Townsend, 1985). In 1986, the book *Girls into Maths Can Go* was published, accompanied by an activities pack (including a video) *Girls into Mathematics*. The book (Burton, 1986) gave an overview of then-current research in gender and mathematics (at that time mainly from North America). The pack (ILEA/OU, 1986) contained activities for teachers to try in their own classroom and to initiate discussion and change with their colleagues. The pack and the video both contained comments and reflections from teachers and pupils providing the reader/viewer with the incentive and confidence to keep on when faced with opposition. "Our aim is to provide you with facts, to help you contend with adverse reactions if and when they occur" (ILEA/OU, 1986). Teacher groups were set up and in-service training was organized around the pack.

Meanwhile, there was a growing understanding of the power of the computer as a tool to enhance the learning of mathematics. The government invested money in curriculum development, and there was an aim of one computer in every mathematics classroom. Some involved in this work were aware of the ways in which girls become alienated from the computers and that computers are nearly always monopolized by the boys (Culley, 1986; Gribbin, 1986; Straker, 1986). Ways of working with the computer in the mathematics classroom were developed that seemed more suited to girls because they developed mathematical understanding through collaboration, discussion, and problem solving (Hoyles, 1988).

The government's Assessment of Performance Unit (APU), which had the task of monitoring the performance in mathematics of large samples of 11- and 16-year-old pupils in England and Wales, provided further data on sex differences in mathematics. One survey pointed out areas of the curriculum where girls were not performing as well as boys. "Boys are best relative to girls in the practical applied areas — measures and rate and ratio — while girls do best relative to boys in computations (whole numbers and decimals) and some aspects of algebra" (APU, 1985). They also found when surveying pupils' attitudes to mathematics that girls expressed greater uncertainty about their mathematical performance than boys. It was also notable that girls underrate their performance and do better in tests than they expect, whereas boys overrate their success and then do less well. (APU, 1988).

In 1986, ten years after Prime Minister Callaghan's statement, the two most important professional bodies in the area, the Royal Society and the Institute of Mathematics and Its Application, issued a report on girls and mathematics (The Royal Society, 1986). The rationale behind the report's commissioning was the awareness that mathematics was acting as a "critical filter to careers" (Burton & Townsend, 1985). The requirement of a good qualification in mathematics at age 16 was keeping girls not only from taking further and higher education in mathematics-related subjects but from entering into careers such as teaching as well.

Government Inspectors Move In

During 1985 to 1988, Her Majesty's Inspectorate (HMI) attempted to throw light on girls' underachievement in mathematics at age 16. Members of the HMI were government inspectors employed (until 1991) to inspect and report on educational institutions. The HMI also produced periodic reviews containing information gained through surveys and school visits, where they went into classrooms and engaged pupils and teachers in discussion. These reviews were independent and sometimes critical of government practice. In their report published in 1989 (HMI, 1989), they were able to pinpoint a set of features that defined a school with successful practices related to gender. They found that these features represented improved teaching for all and that girls in particular benefited from an improvement in the quality of teaching:

It became more and more apparent that girls succeeded in mathematics when teaching was sensitive and perceptive. Naturally in these circumstances boys were equally successful so that in schools where good practice exists neither sex is favoured at the expense of the other. Indeed, focusing on the means of providing girls with opportunities of learning mathematics in congenial and appropriate conditions will help to ensure that all pupils will be able to reach their potential as individuals. (HMI, 1989, p. 29)

The HMI noted that in the first two years of secondary school, pupils were mostly taught in mixed-ability groupings, often using an individualized system of learning. In this system, pupils worked on their own at their own pace. In many of these lessons, there was no teacher exposition or opportunity for discussion and "girls' attitudes and learning suffered disproportionately when exposition and discussion were neglected; they benefited when mathematical work was placed in a relevant context."

The year following the publication of the report, the HMI ran a conference to evaluate how the report's recommendations had been taken up and to identify new directions and actions. In that year, many local education authorities set up working groups of advisors and teachers to act on the report (MacFarlane, 1992; Shropshire Mathematics Centre, 1989).

The Inspectors stated at the end of their report that they "found that focusing on girls' attainment led them into interesting discussions about the nature of mathematics and mathematical learning." Other workers have looked at the nature of mathematics and how beliefs about the subject have led to groups being excluded from its study. For example, Mary Harris in her exhibition "Common Threads" shows how mathematics and textiles are interwoven. "Mary Harris believes that women from a wide variety of cultures should be given credit for the sophisticated level of mathematical thinking that goes into such women's work such as sewing, weaving, knitting and basket work" ("Common threads," 1988).

Plodding or Flair?

The most significant research taking place during this period was at the Girls and Mathematics Unit based at the Institute of Education, London University. During a ten-year period, the Unit conducted studies that analysed the development of girls in mathematics from nursery to the end of secondary school. They analysed video recordings of mathematics classes and exam results from more than 50 London primary and secondary schools. Their results were briefly and strongly put in their report to their funding agency, the Economic and Social Research Council (ESRC, 1988), and more fully expressed by Valerie Walkerdine in her book *Counting Girls Out* (Walkerdine, 1989).

The Unit's findings challenged many accepted notions of the "problem" of girls and mathematics and challenged the myth of girls' failure. Its work shows that even when girls are doing well at mathematics, this is interpreted by teachers as due to "hard work" and "plodding" — attributes seen as not consistent with real mathematical ability:

> ... girls succeed, but their success is achieved not by intellectual ability, rather they are seen as doing well only because they work hard, plod, are unproblematic pupils in that they can get on with their work quietly and are not disruptive in the class-room. 'Hard work' becomes a pejorative attribution (especially in middle class schools), used to denote the very crucial absence of 'real' ability, even though hard-working girls are typically viewed positively as pupils. (ESRC, 1988, p. 4)

It seems that girls can never be really good at mathematics. Boys on the other hand are seen to have "flair," "real ability," and "brilliance." For a boy to be bored or disruptive is often interpreted by the teacher as a sign of intelligence. The Unit quoted one teacher it interviewed to say:

> I would say he is the brightest child in the whole class but again he is someone who needs to be encouraged, he can be very rude, and can be obstreperous with other children rather than the teacher; but he has a very good ability, very interested in everything, a very good general knowledge, and has an all round ability with a lot of potential. (ESRC, 1988, p. 5)

What do girls have to do in order to be taken seriously in mathematics by their teachers? The report says that "teachers must examine their judge-ments and monitor their reactions to help girls fight back."

A NEW EXAM HELPS GIRLS

In the 1980s, gender was firmly on the agenda, and the performance of girls in secondary school was improving. This had still failed to have a great impact on girls' performance in exams. This was changed by the introduc-tion in 1988 of the GCSE (General Certificate of Secondary Education) for children at age 16. In mathematics, GCSE represented a radical change from the old examination system. It adopted differentiated assessment tech-niques following the underlying philosophy of the Cockcroft report, which had said:

> We believe that there are two fundamental principles which should govern any examination in mathematics. The first is that the examination papers and other meth-ods of assessment which are used should be such that they enable candidates to demonstrate what they do know rather than what they do not know. The second is that examinations should not undermine the confidence of those who attempt them. (Cockcroft, 1982, para. 521)

As well as traditional exam papers, the GCSE in mathematics was to include some coursework assessment as a way of allowing pupils to show what they do know. And it was a unified examination, replacing the divi-sion into O-level or CSE, which did not undermine confidence in the way that the divided examinations did. It was predicted that both of these would prove beneficial to girls, and this proved to be the case.

In 1991, when GCSE had been in place for three years, the Schools Examinations and Assessment Council (SEAC) commissioned an investigation into gender differences in GCSE English and mathematics. The report, published in 1992, found that "the gender differences in performance in GCSE mathematics which have traditionally favored boys are steadily decreasing and have narrowed to the extent that boys gain only 3% more grades A–C than girls" (Stobart, Elwood, & Haydon, 1992a, p. 3). Girls may be doing even better relative to boys than was shown by the statistics, because a lower proportion of boys are entered, which could mean that low attaining boys are not entered. The data for the years 1988 to 1991 are show in Figures 1 and 2.

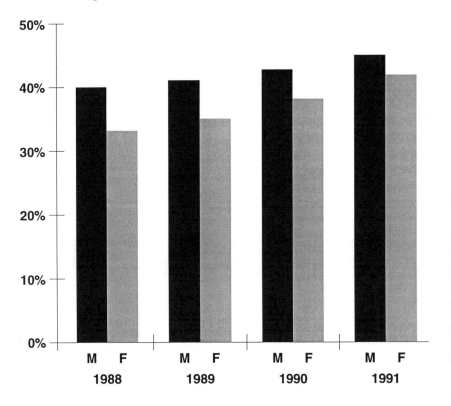

Figure 1: Percentage of pupils obtaining GCSE maths grade A–C,
by gender

Girls Gain From Coursework

The new GCSE meant classrooms where "mathematics for all pupils is to be understood as having a creative and imaginative dimension" and "collaborative work, experimental and extended pieces of work, both practical and investigative, are legitimate forms of activity in school mathematics" (Isaacson, 1987).

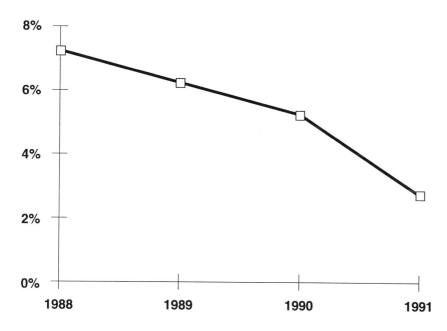

Figure 2: GCSE maths grades A–C; declining difference
between boys and girls

Cockcroft had noted that "if assessment at 16+ is to reflect as many aspects of mathematical attainment as possible, it needs to take account not only of those aspects which it is possible to examine by means of written papers but also of those aspects which need to be assessed in some other way" (para. 534) Nevertheless, coursework was a new departure in mathematics assessment.

Although girls are still 3 percent behind boys in the GCSE, they perform better than boys on mathematics coursework, which was encouraged by the new GCSE. Maria Goulding ("Let's hear it," 1992) examined the data from one examination board for 1991, the year when coursework became compulsory in mathematics. There were data from four different syllabuses with amounts of coursework varying from 20 percent to 50 percent of the assessment marks. She found that girls outnumbered boys in the higher grades (A–C) in the two options where examinations only counted for 50 percent of the final marks.

Single Exam

The introduction of GCSE was also welcomed as it provided a single examination. Pupils at 16 would no longer be classified as O-level (high attaining) or CSE (average or low attaining). The Girls and Mathematics Unit had found that girls were more likely to be entered for the CSE examination, even when they perform better than boys who were considered suitable for O-level. "What is most important and alarming is that despite the fact that girls performed overall better than boys in four out of five classes, far fewer girls than boys were entered for 'O' level mathematics" (ESRC, 1988). Although a grade 1, the highest grade at CSE, was equivalent to a grade C in O-level, it was in practice often not accepted as such. Entry to A-level mathematics and to some statistics-related courses at university required what was called a "real" grade C.

However, the downgrading of girls' attainment did not stop with the change to a new examination. In mathematics at GCSE, pupils are entered for one of three tiers with restricted grade ranges: higher tier (grades A–D), intermediate (grades C–F), or foundation (grades E–G). A pupil entered for a particular tier whose grade falls outside the range of that tier is awarded an unclassified. A grade A to C is needed for entry into many careers (including initial teacher education), which means pupils entered for the intermediate must gain the top grade. "The larger entry of females into the intermediate tier has been a common feature of GCSE mathematics since its introduction in 1988" (Stobart, Elwood, & Haydon, 1992). It seems that this high proportion are entered for the middle level as the safety option which provides the key grade C whilst avoiding the risks of being unclassified. "There was evidence of at least grade B level performance for those at the top of the Intermediate tier where the highest grade possible is a grade C" (Stobart, Elwood, & Quinlan, 1992).

It seems teachers choose the safety option for girls to compensate for their perceived lack of confidence. Teachers explained their reasons:

- the weaker girls feel more secure
- a tendency (of girls) to lower expectation of self
- boys tend to push for higher level entry rather than intermediate ... girls tend to push for intermediate level entry and ... are not so concerned about higher level entry. (Stobart, Elwood, & Haydon, 1992, p. 30)

Researchers looking at the introduction of GCSE in Northern Ireland found the same thing (Evason, Lawlor, & Morgan, 1990). This underestimation of what girls can do is preventing some from continuing with their mathematics at A-level. Some schools require a grade B for entry to A-level schools while others will only take a grade C from the higher tier. "This therefore marginalises the candidature taking the Intermediate tier from taking their mathematics further, especially the disproportionate number of girls who are entered for this tier" (Stobart, Elwood, & Haydon, 1992).

Fewer Women at Higher Levels

More girls than boys are taking GCSE mathematics, and almost equal proportions of them are gaining the higher grades A to C, yet fewer of these girls continue the study of mathematics to A-level. In 1992, 35 percent of the 18-year olds attempting A-level mathematics were girls and of these 78.3 percent gained a pass compared with 77.2 percent of the boys (Statistical Bulletin, 15/93). This represents a steady but small improvement over the years shown in Tables 1 and 2. In 1983, 31.8 percent of the entry were girls, and for both boys and girls the pass rate was 74.5 percent. In 1979, 73.3 percent of the girls had passed mathematics and they had made up 28.6 percent of the entry. At A-level, the pass rate in mathematics is increasing and girls are leading.

Table 1: A-level entries for mathematics

England and Wales								
1992	1991	1990	1989	1988	1987	1986	1983	1979
Girls	35.0%	35.4%	36.0%	34.9%	33.5%	32.4%	32.3%	31.8%	28.6%
Boys	65.0%	64.6%	64.0%	65.1%	66.5%	67.6%	67.7%	68.2%	71.4%

Table 2: Percentage of girls and boys passing A-level mathematics

England and Wales								
1992	1991	1990	1989	1988	1987	1986	1983	1979
Girls	78.3%	79.3%	78.7%	78.5%	75.5%	75.5%	73.5%	74.5%	73.0%
Boys	77.2%	77.2%	76.6%	76.1%	77.7%	75.2%	76.7%	74.5%	73.4%

(Source: *Department for Education School Leavers Survey*)

The percentage of women studying mathematics decreases as the level increases. At UK (including Scotland and Northern Ireland) universities in 1992/93, only 25.8 percent of undergraduates studying mathematical sciences were women (University Funding Council, 1993). This falls to one in five for postgraduate degree courses. From 1984, there have been many initiatives to widen access and encourage women into science and engineering through women-only access courses. However, the women were sought not from a desire for greater equality but for their numbers. Tessa Lovell and Mal Leicester said: "We discovered good practice in recruitment of female students, particularly within Physics departments. Yet, this was not indicative of good practice across the spectrum. Women were perceived as an obvious means of generating more students" (Lovell & Leicester, 1993).

There has been no observable commitment to change to the composition of the university mathematics department. In 1992/93 (University Funding Council, 1993) in British (including Scottish) universities, there were five women professors of mathematics compared with 256 men (1.9%), 13 out of the 403 readers and senior lecturers, and 89 out of 777 lecturers. These numbers have remained relatively unchanged since 1987 when data was first published giving a breakdown according to gender (see Table 3).

Table 5: Full time academic staff in university mathematics departments

	Great Britain				
	Professors		Readers/senior lecturers		Mean age
	Men	Women	Men	Women	(years)
1992–93	256	5	390	13	45.9
1991–92	248	3	378	13	45.6
1989–90	222	2	362	11	45.4
1987–88	219	2	370	9	45.0
	Lecturers		Other grades		Mean age
	Men	Women	Men	Women	(years)
1992–93	688	89	8	2	45.9
1991–92	690	82	3	2	45.6
1989–90	740	73	3	6	45.4
1987–88	822	69	4	1	45.0

(Source: *University Statistics: Volume 1 – Students and Staff*)

The average age of university staff has been steadily increasing each year, which illustrates that there are very few young staff gaining positions. Unless there is a radical change, university mathematics departments will continue to be totally male at the top.

1988 TO 1994: TAKING AWAY GAINS GIRLS HAD MADE

By the late 1980s, just as girls had gained recognition for their mathematics ability and were reaping the benefits of changes in the education system, there was a profound and reactionary transformation in British education. In six years, all the agencies promoting equality were curbed or disbanded; a curriculum that pays no service to equal opportunities is being institutionalized, and testing improvements have been withdrawn. Gender issues are no longer considered important; girls' performance is expected to fall again.

Why has this happened? Who has forced these changes? Professor Margaret Brown in her 1993 inaugural lecture as professor of mathematics education at King's College, London, sees a "battle for control" taking place:

> Educational policy in relation to schools and teacher training in England has for the last six years been significantly affected by the views of a clique of not more than a dozen people on the radical right, unaccountable and unelected, based in and around universities, with knowledge of neither educational research nor educational theory, and, most important with little or no experience of working within the state educational system in this country. (Brown, 1993, p. 98)

Members of this group are based around the Centre for Policy Studies and have been appointed to key positions in committees and advisory groups dealing with the curriculum.

> Amazingly, considering the smallness of the group in contrast to the number of those engaged in education, they have been, and to some extent still are, strongly represented on every working group, committee and quango connected with education. (Brown, 1993, p. 98)

This group, rather than the government, has become the driving force behind changes in education. In part, this group became powerful because, in education, the government has been rudderless, with five education ministers in a six-year period. This group has an agenda in education to return to what it calls "traditional values"; it sees mathematics as consisting entirely of facts, skills, and techniques. The radical idea that mathematics can be learned by discussion, practical and investigational work — ideas made respectable by the Cockcroft Committee — were questioned. "There is no reason to imagine that pupils learn from talking.... The acquisition of knowledge requires effort and concentration," wrote Sheila Lawlor of the Centre for Policy Studies (quoted in Brown, 1993, p. 102).

Curbing Progressive Agencies

In 1988, the government set out to break the power of progressive local education authorities (LEAs). LEAs have the responsibility for all education in their area. The Right targeted the largest and most progressive, the Inner London Education Authority (ILEA). In 1987, the radical Right, then in a body called the "Hillgate Group," stated that

> The ILEA ... has tolerated or even encouraged a politicised curriculum and the introduction of politicised appointments and teachers; finally, the entrenched communist and neo-communist domination of the Inner London Teachers' Association, with consequent continuing high level of militant activity, obliges a democratic government to do what it can to destroy that Association's power-base. (Hillgate Group, 1987)

ILEA was abolished the following year, and its work was distributed to 12 new inner-London LEAs. Good practices promoted by the ILEA that were anti-sexist and anti-racist were not necessarily continued and monitored by these boroughs. Other forward-thinking LEAs, such as Brent, were rendered powerless by the government's encouraging schools to "opt out" of their control to become self-managed. When a school opts out, it is allocated a larger than normal share of the LEA budget, which means that the remaining schools suffer. Central services, including advisors and advisory teachers, then need to be cut, particularly when several schools opt out. Advisors and advisory teachers were key in promoting and supporting changes in the classroom, particularly around equality issues, and their numbers have been reduced. "There is evidence of less INSET (In-service training) devoted to secondary mathematics. This stems from changes in priorities in schools and from changes in job specifications for many mathematics advisors which has reduced the time they can devote to the subject. There are also fewer advisory teachers" (Office for Standards in Education, 1992).

In 1991, the school's inspectorate service was cut back. Her Majesty's Inspectorate staff was reduced to 40, and a new government agency, Office for Standards in Education (Ofsted), was set up. The task of the 40 remaining HMIs was to set up, train, and monitor private inspection teams. According to Education Minister Kenneth Clarke, the reform was to introduce "diversity and competition" ("Power to the people's HMI," 1991). In 1989, the HMI published the report *Girls and Mathematics*, and this was followed in 1990 with a conference to set guidelines for action still to be taken. With the effective end of the HMI, there is no evidence that any action suggested at this conference will be taken.

The Assessment of Performance Unit was the only government institution with the task to survey the performance of girls and boys in mathematics. In 1989, it, too, was closed down by the government. There is a new unit in the Schools Examination and Assessment Council (SEAC) with responsibility for monitoring and evaluating the national curriculum assessment arrangements. It is not clear whether gender differences will be part of its brief.

The Girls and Mathematics Unit finished its work in 1988, although its findings stayed in the public eye for some years, following the publication of its work in the book *Counting Girls Out* (Walkerdine, 1989). Since then, research into gender and mathematics has fallen off the agenda. A look at research journals and conference proceedings in mathematics education in Britain shows that mainstream investigations into mathematics education are not adopting a gender perspective. An area where the gender differences are wide and widening is that of information technology. Some research on the use of computers for the learning of mathematics is well informed by the gender dimension. However, at a large international conference on technology and mathematics teaching (TMT, Birmingham 93), there were no research papers that focussed on issues of gender.

Education Reform Act Is the Final Blow

The Education Reform Act (ERA) became law in England and Wales in 1988. It brought about some of the greatest changes to education for 50 years. Its key ingredient was the imposition of a national curriculum. The national curriculum was heralded by the government as a curriculum that gives entitlement to all, because all pupils are treated the same and all pupils will be entitled to receive a balanced, broadly based, and relevant curriculum, regardless of sex, ethnic origin, and geographic location. Working groups were set up to develop the "balanced" curriculum. These groups were not balanced. They had a majority of men, with no members of ethnic-minority communities. For this and other reasons, many writers have questioned the so-called equality of the national curriculum (Burton & Weiner, 1990; Carr, 1989; Hoyles, 1990; Kant, 1987; Roberts, 1987). Kant says the question about the curriculum is: "Good for whom and relevant for what?" Arnot answers that a curriculum "based on white, Anglo-Saxon, Christian cultural heritage, and not valuing the contribution of black and ethnic minority cultures, languages and religions" is not particularly "good or relevant" for many groups of pupils (Arnot, 1989).

Also, the ERA can do nothing to change attitudes or stereotyping. Linda Carr states:

> Low expectations and negative attitudes are deeply damaging — possibly to black girls and those from minority ethnic groups most of all — but are very difficult to quantify. The ERA, like the SDA [Sex Discrimination Act] and other legislation, can do nothing about attitude and it will be interesting to see whether, for example, girls' performance in maths, which is and has been a core subject, will improve as a result of the National Curriculum. (Carr, 1989)

All agree that the ERA was not about creating equality but about the government taking control over what was happening in the classroom. When the ERA was proposed, there was a classroom climate of curriculum development and innovation. Teachers could design their own curriculum, and some were choosing a curriculum that aimed to be anti-racist and anti-sexist. Such freedom was worrying to Kenneth Baker (the Education Minister in 1987) who stated: "We cannot continue with a system under which teachers decide what pupils should learn without reference to clear nationally agreed objectives and without having to justify their decisions to parents, employers and the public" (Roberts, 1987). Arnot noted that "the increasing political control of education has effectively narrowed the spaces in which anti-sexist and anti-racist initiatives and policies can operate" (Arnot, 1989).

DID GIRLS DO TOO WELL?

Testing became a focal point of the gender battle, with the new Right equating better performance by girls with falling standards. With the change to GCSE in 1988, girls started to improve their performance: "The gap in mathematics is considerably smaller and narrowing" (Stobart, Elwood, & Haydon, 1992). The introduction of coursework was a significant factor in girls' improvement, not just in mathematics but in all subjects. But this high performance of girls was suspect; Education Minister Kenneth Clarke talked about the "grade drift up" in English, where in many cases, assessment is by coursework only. Clarke said he thought "it unlikely that this increase was merely due to an increase in standards" ("GCSE scores," 1991).

Under the new national curriculum, all pupils in state schools are to be assessed against attainment targets at four "key stages" at ages 7, 11, 14, and 16 years. The first pilot tests were in 1990, and they were for 14-year olds at "key stage 3"; these tests assessed pupils' mathematical attainment by "standard assessment tasks" (SATs), which were practical and open ended. Each task covered a range of attainment targets across all levels. Each was designed to take between one and three weeks and fit into the pupils' normal school week.

Girls performed better than the boys on these open-ended tasks in the pilot test — in sharp contrast to all other tests for this age group. "These data show that the attainment of the girls was in general higher than that of the boys, with the mean score of the girls exceeding that of the boys in 3 of the 4 SATs and in PC1 [number and algebra] of the fourth SAT" (Consortium for Assessment and Testing in Schools, 1990).

The following year, the government changed the form of the SATs to a series of short, one-hour, paper-and-pencil tests — a form of assessment that girls traditionally do less well in. The implication was obvious even to the mainstream educational press. A *Times Educational Supplement* headline read: "Test changes threatens girls: The government's assessment advisers are warning that girls could lose out as a result of the switch to short written test for 14-year-olds" ("Test change," 1991).

The official reason given for the change was to reduce the time load on teachers; there had been substantial teacher protest about the tests, and the change did reduce the time needed. But in claiming to respond to teacher complaints, the government took out the only aspects that teachers found valuable; subsequent tests were boycotted by teachers. The whole issue of national-curriculum testing arose in a very contradictory environment. On the one hand, teachers were strongly opposed to the tests because they felt that the tests were designed for ranking rather than diagnostic purposes, and that they would have to teach to the test rather than teach mathematics (Goldstein, 1990; Hoyles, 1990). On the other hand, teachers discovered that certain aspects of the pilot tests were useful and that "the tasks gave opportunity for learning as well as for assessment" (Consortium for Assessment and Testing in Schools, 1990). Teachers reported that group work improved the pupils' enjoyment and, more important, that group work gave the pupils opportunity to gain communication skills: Some pupils were able to work at a higher level than predicted through being part of the group. A high proportion of pupils (85–97%) said that they enjoyed some or all of the activity.

Cutting Down Coursework

Just five months before the decision to take open-ended tasks out of the national curriculum exam, the government announced a change which had a similar effect — the proportion of coursework allowed in GCSE was to be cut to 20 percent. This change was announced by the Prime Minister, John Major, in July 1991 at a meeting hosted by the Centre for Policy Studies, the home of the radical Right referred to earlier by Professor Margaret Brown. The decision was backed by the Education Minister, who said that HMI supported the reduction, even though at a previous meeting held to evaluate the GCSE, both HMI and head teachers had been favorable to GCSE. "HMI reported positive improvements in pupil performance and achievement and in teaching quality [and] head teachers say the GCSE has had a positive impact on 16-plus staying-on rates" ("Blow your own trumpet," 1988). Later, the decision was revised. Only the mathematics GCSE was to be limited to 20 percent coursework; other subjects still had to reduce coursework, but could maintain levels above 20 percent. This revision was as a result of advice given by The Schools Examination and Assessment Council. However, this advice was strongly disputed in a letter from the members of the SEAC mathematics committee. In the letter, the committee — made up of teachers, lecturers, and mathematics advisors — said that their advice had not been sought by SEAC. If consulted, they would have advised that "maths GCSE contain up to 70% coursework" ("Maths GCSE," 1991). They added that they were in favor of retaining coursework, as "coursework assessment allows pupils to demonstrate what they know and has led to an improvement of standards.... Girls, in particular, seem to do better at maths when motivated and challenged by coursework." Two members of the committee did not sign the letter. One of them was Dr. John Marks, who was chair of the committee and associated with the Centre for Policy Studies, referred to earlier.

A-Level Distinctions

The small improvement by older girls — at A-level — was also challenged and treated with suspicion. As noted before, the pass-rate in A-level mathematics is increasing for boys and girls. But in 1991, a higher percentage of girls than boys passed, 79.3 percent compared 77.3 percent. Of these, 18 percent of girls compared with 21 percent of boys gained an A, the highest grade. In 1993, the Department for Education set up a consultation process to look at A-level results. Education Minister John Pattern said he wished to find a way to recognize more effectively A-level candidates of exceptional ability. A consultation document (*Consultation on Starred A Grade and AS Examinations*, issued by the School Curriculum and Assessment Authority) said the inclusion of a new grade A* ("A-star") was needed as the "existing A level scale does not offer sufficient reward for those candidates who do exceptionally well." Also, "increasing numbers of candidates have been gaining A level qualifications. The introduction of a new grade will help reassure the public that standards are being maintained."

But which students are of "exceptional ability"? Is the grade A* needed to separate out "exceptionally able" boys from the "hard-working" girls? According to the work of the Girls and Maths Unit, girls performance is seen as due to hard work and effort rather than exceptional ability. Girls are not considered to have "flair and brilliance" (ESRC, 1988).

Curriculum Changes Hurt Girls

Aside from the move to pencil-and-paper tests from open-ended tasks, there have been continuous changes to the national curriculum itself — making it very different from that first produced by the early working parties and removing those changes which had most benefited girls. In August 1988, the Mathematics Working Party published its report. One important aspect of this report was that assessment should be based on three "profile components." Profile Component 3 (PC3) was "practical applications of mathematics in terms of personal qualities, communication skills, and using mathematics." The government would not allow "personal qualities" to be assessed in mathematics and put pressure on the National Curriculum Council (NCC). When the new curriculum was published in December 1988, it contained only two profile components. The third, PC3, had been integrated into the other two and the terms "personal qualities" and "communication skills" had been removed. This was after the NCC noted that over 80 percent of the people consulted voted to maintain PC3. Duncan Graham, Chair of the National Curriculum Council, was blamed for succumbing to government pressure to lose PC3. "Although he revamped the rest of the report to take in some of the maths working group's recommendations, it was generally felt that he had toed the Government line and demoted practical applications below basic skills and rote learning" ("NCC Chairman says," 1989).

Following a large-scale review and consultation, proposals for a new "slimmed-down" national curriculum were published on May 9, 1994. An advisory group for mathematics had worked to produce the document. But, unexpectedly, an alternative from the new Right was also tabled, provoking an unprecedented press statement from the Joint Mathematical Council:

> All subjects contain one set of proposals with the exception of Mathematics, which has an alternative proposal in which the attainment target relating to "Using and Applying Mathematics" is integrated into the remaining content targets (Number and Algebra; Shape, Space & Measurement; Handling Data). The JMC believes that this

alternative was not proposed by the mathematics committee set up by the Schools Curriculum and Assessment Authority, but has been included at the personal request of a member of the Authority, John Marks, with the support of the Chairman, Sir Ron Dearing, and the Chief Executive, Chris Woodhead, not the full Authority. (Press release by The Joint Mathematical Council of the UK)

The alternative proposal set out to finally get rid of the practical applications of mathematics. Each change to the curriculum diminished the role of "Using and Applying Mathematics." Its presence allows some opportunity to bring issues of equality into the maths classroom. "It could be suggested that the inclusion of this statement makes possible the implementation of a processed-based approach to teaching mathematics: the kind of approach that would encourage styles of learning and teaching, and curriculum choices that support the elimination of racism and sexism" (Burton & Weiner, 1990). The government cannot allow such styles of learning and teaching and has permitted one member of the authority, John Marks, to countermand the advice and views of the mathematics committee. Do girls have any hope of being allowed to succeed?

REFLECTIONS

In the 1980s, girls did gain a voice in the mathematics classroom — even if they sometimes had to shout. Why was this allowed under several years of Conservative rule and then suddenly reversed? Madeleine Arnot suggested that government action on equal opportunities was not about creating justice and social equality or about empowering women, but rather was "part of an economic strategy in which greater, though limited, investment in women [was] thought useful as a source of skilled labour, particularly as scientists and engineers" (Arnot, 1987). In the early 1980s, the shortage of well-trained scientists and engineers was seen as so acute that women-only conversion courses were set for women with non-science backgrounds. Because of this need to ensure a supply of well-trained scientists, teachers' efforts to develop an anti-sexist and anti-racist curriculum was legitimized and even became part of the mainstream mathematics curriculum in schemes such as SMILE (The School Mathematics Individualised Learning Experience).

Five years later, the women who were the first to benefit from these conversion courses (described in Isaacson & Smart, 1988) also became the first casualties. When they graduated, there were few jobs and women were no longer welcome in the male workplace.

The fate of these women provides one explanation for the Conservative Party's change of heart. In Britain there is an economic recession and there is no longer a need to encourage women in mathematics and science. This has allowed middle- and upper-class white men to reassert their dominance and take away the gains conceded to girls and women in the 1980s. What is most striking is the speed with which this was done and the total inability of the educational establishment — and of women — to resist it.

REFERENCES

Arnot, M. (1985). *Race and gender: Equal opportunities in education.* Oxford: Pergamon Press & The Open University.

Arnot, M. (1987). Political lipservice or radical reform: Central government responses to issues of sex equality. In M. Arnot & G. Weiner (Eds.), *Gender and the politics of schooling.* London: Hutchinson.

Arnot, M. (1989). Crisis or challenge: Equal opportunities and the national curriculum. *NUT Education Review, 3*(2), 7–13.

Arnot, M., & Weiner, G. (1987). *Gender and the politics of schooling.* London: Hutchinson.

Assessment of Performance Unit. (1985). A review of monitoring in mathematics, 1978 to 1982. London: HMSO.

Assessment of Performance Unit. (1988). Attitudes and gender differences. Windsor: NFER-Nelson.

Blow your own trumpet, examiners told. (1988, September 30). *Times Educational Supplement.*

Brown, M. (1993). Clashing epistemologies: The battle for control of the national curriculum and its assessment. Inaugural Lecture given by Margaret Brown, Professor of Mathematics Education, at King's College London. *Teaching Mathematics and Its Applications, 12*(3), 97–112

Burton, L. (1986). *Girls into maths can go.* London: Holt Education.

Burton, L., & Townsend, R. (1985, March). Girl-friendly mathematics. *Mathematics Teaching.*

Burton, L., & Weiner, G. (1990). Social justice and the national curriculum. *Research Papers in Education, 5*(3), 203–227.

Carr, L. (1989). Equal opportunities — Policy and legislation after ERA. *NUT Education Review, 3*(2), 19–23.

Clarricoates, K. (1987). Dinosaurs in the classroom — The "hidden" curriculum in primary schools. In M. Arnot & G. Weiner (Eds.), *Gender and the politics of schooling.* London: Hutchinson.

Cockcroft, W. H. (1982). *Mathematics counts.* London: DES.

Common threads. (1988, March 4). *Times Educational Supplement.*

Consortium for Assessment and Testing in Schools. (1990). CATS Mathematics Trial, interim and final report. London: Kings College.

Culley, L. (1986). *Gender differences in secondary schools.* Loughborough: Department of Education, Loughborough University.

DES. (1985). *Better schools.* London: HMSO.

Drew, D., & Gray, J. (1990). The fifth-year examination achievements of black young people in England and Wales. *Educational Research, 31*(2), 107–117.

ESRC. (1988). *Girls and Mathematics: Some lessons for the classroom.* London: The Girls and Mathematics Unit.

Evason, E., Lawlor, S., & Morgan, V. (1990). GCSE mathematics the first year's experience: Possible gender differences in response to the new 16+ examinations. *CORE, 14*(2), 1 (microfiche).

GCSE scores to hang on final exam. (1991, November 22). *Times Educational Supplement*.

Goldstein, H. (1990). The fundamental assumptions of national assessment. In P. Dowling & R. Noss (Eds.), *Mathematics versus the national curriculum*. London: Falmer Press. .

Gribbin, M. (1986). Boys muscle in on the keyboard, in L. Burton (Ed). *Girls into maths can go*. London: Holt Education.

Headlam Wells, J. (1985). 'Humberside goes neuter': An example of an LEA intervention for equal opportunities. In J. Whyte, R. Deem, L. Kant, & M. Cruickshank (Eds.), *Girl friendly schooling*. London: Methuen.

Her Majesty's Inspectorate (HMI). (1989). *Girls Learning Mathematics*. London: DES.

Hillgate Group. (1987). *The reform of British education*. London: The Claridge Press.

Hoyles, C. (1988). *Girls and computers*. London: Institute of Education.

Hoyles, C. (1990). Neglected voices: Pupils' mathematics in the national curriculum. In: P. Dowling & R. Noss (Eds), *Mathematics versus the national curriculum*. London: Falmer Press.

ILEA/OU. (1986). *Girls into mathematics*. Milton Keynes: Open University.

Isaacson, Z. (1987). *Teaching GCSE mathematics*. London: Hodder and Stoughton.

Isaacson, Z., & Smart, T. (1988). It was nice being able to share ideas: Women learning mathematics. *IOWME Newsletter, 4*(2).

Joseph, G. G. (1991). *The crest of the peacock*. London: I. B. Tauris & Co.

Kant, L. (1987). National curriculum: Notionally equal? *NUT Education Review, 1*(2), 41–49.

Kelly, A. (1981). *The missing half: Girls and science education*. Manchester: Manchester University Press.

Let's hear it for the girls. (1992, February 21). *Times Educational Supplement*.

Lovell, T., & Leicester, M. (1993). Gender and higher education. *Journal of Further and Higher Education, 17*(1), 52–56.

MacFarlane, K. (1992). Mathematics. An LEA based INSET initiative. *Mathematics in School, 21*(3), 34–35.

Maths GCSE snub infuriates experts. (1991, February 6). *Times Educational Supplement*.

NCC chairman says maths topic will be restored. (1989, September 22). *Times Educational Supplement*.

Office for Standards in Education. (1992). *Mathematics: Key Stage 1, 2, 3*. London: Office of HM Chief Inspector of Schools.

Plewis, I. (1991). Pupils' progress in reading and mathematics during primary school: Associations with ethnic group and sex. *Educational Research, 33*(2).

Power to the people's HMI. (1991, November 8). *Times Educational Supplement*.

Roberts, A. (1987). Multi-culturalism and anti-racism: Where is the leadership? *NUT Education Review, 1*(2), 45–49.

Shan, S., & Bailey, P. (1991). *Multiple factors: Classroom mathematics for equality and justice*. Stoke-on-Trent: Trentham Books.

Shropshire Mathematics Centre. (1989). *Girls learning mathematics in Shropshire.* Shropshire Mathematics Centre.

Statistical Bulletin 15/93. (1993). *School examination survey 1991/92.* London: DFE.

Stobart, G., Elwood, J., & Haydon, M. (1992). *Differential performance in examinations at 16+: English and mathematics.* London: ULEAC and NFER.

Stobart, G., Elwood, J., & Quinlan, M. (1992). Gender bias in examinations: how equal are the opportunities? *British Education Research Journal, 18*(3), 261–276.

Straker, A. (1986). Should Mary have a little computer? In L. Burton (Ed), *Girls into maths can go.* London: Holt Education.

Taylor, H. (1985). A local authority initiative on equal opportunities. In M. Arnot (Ed.), *Race and gender: Equal opportunities policies in education.* Oxford: Pergamon Press & The Open University.

Test change threatens girls. (1991, November 29). *Times Educational Supplement.*

Tests tried and found wanting.(1994, July 8). *Times Educational Supplement.*

The Royal Society and The Institute of Mathematics and Its Applications. (1986). *Girls and mathematics.* London: The Royal Society.

University Funding Council. (1993). *University statistics 1992–93: Vol. 1 students and staff.* London: Universities Statistical Record.

Walkerdine, V. (1989). *Counting girls out.* London: Virago.

Weiner, G., & Arnot, M. (1987). Teachers and gender politics. In M. Arnot & G. Weiner (Eds.), *Gender and the politics of schooling.* London: Hutchinson.

Whyte, J. (1986). *Girls into science and technology.* London: Routledge & Kegan Paul.

Whyte, J., Deem, R., Kant, L., & Cruickshank, M. (1985). *Girl friendly schooling.* London: Methuen.

JANICE M. GAFFNEY

AND

JUDITH GILL

GENDER AND MATHEMATICS IN THE CONTEXT OF AUSTRALIAN EDUCATION

In Australia, school mathematics education must be understood in terms of both the provisions associated with the different states and the schooling structures nationally. In each of the six states, there is a strong tradition of state ownership and control over school structures; the curriculum in the senior years is prescribed by the state authority. However there is considerable difference between the various authorities in terms of the knowledge seen as appropriate for different subjects and the ways in which the knowledge is divided up and assessed. For instance, the Year 12 mathematics curriculum in Tasmania could involve quite different concepts and applications as compared to the mathematics curriculum in, say, Queensland. In each case, the local universities are represented on the curriculum committees and have a good deal of influence in determining the curriculum. Given a high degree of population stability within the Australian states and within Australian universities, mathematics teachers tend to have studied at the local university and to have taken on the particular predilections of that mathematics department. The outcome is that there are real and important differences in mathematics content between the states and associated differences in subject groupings and assessments, a feature which makes comparison between the states difficult and generalization across the country highly problematic. Currently the federal government is pushing for the adoption of national curriculum statements and profiles relating to all subject areas. While this move does not equate to the prescription of curriculum content, it will be interesting to monitor the effect of even this degree of nationalization on the currently disparate mathematics curriculums across the states.

NATIONAL SCHOOLING STRUCTURE

In Australia generally, compulsory schooling finishes at age 15. Historically the majority of the student group left school at that age, having completed approximately ten years of schooling and usually reaching the third-last year of secondary school. Retention rates have increased markedly in the past 10 to 15 years, a situation which has led to a renewed and vigorous interest in

G. Hanna (ed.), Towards Gender Equity in Mathematics Education, 237–255.
© *1996 Kluwer Academic Publishers. Printed in the Netherlands.*

the "post-compulsory" senior years. Young people have been staying on at school rather than join the lines at the employment offices, a feature which has produced a re-examination of the senior school curriculum. Traditionally, Australian schools in their senior years offered a curriculum that was directed specifically at the academic requirements of universities, as it was assumed that those students who completed schooling were bound for university. In the last decade, the Australian government has initiated significant changes in the area of higher education in an effort to change the elitist nature of the old universities. More young Australians are proceeding to tertiary education in recent years as a direct result of the government's restructuring of the tertiary sector. Consequently, a broader cross section of the senior school population aspires to "go to uni," an outcome consistent with the broader definition of what "uni" means and what courses are available there. Because of the changes in the senior school population noted above, the curriculum in this part of the school has now been considerably broadened to incorporate areas considered more relevant to the needs of young people who are not necessarily going on to further study or who are going on in a broader range of areas than was traditionally the case. The older connection between success in senior school and traditional university courses lives on, however, and nowhere more surely than in the mathematics curriculum. The "maths/science" strand is unofficially, and sometimes officially, known as the "hard" subjects, undertaken by "bright" and/or highly ambitious students, who are seen by their teachers, parents, and themselves as destined for courses at traditional universities. Typically, mathematics teachers are keen to encourage students who appear capable in these areas to continue with them. Teachers in other subject areas complain that bright students are made to feel that enrolment in anything other than mathematics/science is inappropriate. Australian students typically study five subjects in their final year of school, with some states making English a compulsory requirement. The choices made by students are thus fairly restricted, although less tightly than, for example, in the United Kingdom. The connection between high ability and participation in mathematics at senior school is continually reinforced by student and teacher perception, interrelated with the structuring of subjects within the curriculum. The impact of this practice on gender division in subject participation and related achievement will be taken up later in this chapter.

MATHEMATICS EDUCATION IN AUSTRALIAN SCHOOLS

Despite the concerns raised earlier about differences in mathematics education between states, there are some national generalizations that can be made. All students in Australian schools undertake mathematics as a normal and important subject within the primary school curriculum. Mathematics continues in all states as part of the core curriculum during the compulsory school years, although there is some variation between and within states in terms of how much and which mathematics is studied at different levels. By the senior post-compulsory years in all states there operates what can loosely be described as a triple-banded tracking structure in the senior mathematics curriculum (after Ainley, Jones, & Navaratnam, 1990; Jones, 1986; Rosier, 1980). In other words, the students who undertake mathematics in the senior school are grouped in one of the following streams.

Band 1 is the old A stream. It consists of advanced or professional double mathematics courses — known unofficially as "the hard maths" — which lead to the study of higher mathematics at university or, alternately, to engineering or science degrees. In 1990, 12.6 percent of students were in this band (plus 4.6% for the New South Wales students in the 1a or inter-advanced band); in 1991 the situation was much the same with roughly one student in eight undertaking advanced mathematics nationally (Ainley, Jones, & Navaratnam, 1990; ESSA, 1993). Students in this mathematics stream tend to be represented among the highest achievers in terms of tertiary entrance scores. (The TE score or tertiary entry score determines which courses students may enrol in at university. It is calculated by aggregating scores gained in final year subjects. In some states, there is a scaling process calculated on the basis of subject groupings, in others not. Different states use different combination of subjects from which to calculate the TE score; in some cases it is a simple aggregate of all scores, in others it is the aggregate of the top three or four with percentage points added for extra subjects. Students base senior school enrolment decisions on their knowledge of, among other things, how these calculations operate in their state.)

Band 2 is a fully accredited senior mathematics course which usually involves fewer class hours than Band 1 and counts as one unit rather than two. Enrolment in this stream is also associated with high achievement in terms of TE scores, but less strongly than with Band 1. Students who take this stream are aiming for university but not for higher mathematics or engineering; their orientation may be spread across the range of desirable courses from medicine through architecture, journalism, and so on, in other words courses which require little or no formal mathematical knowledge. In 1990 34 percent of senior students chose this band of mathematics (Ainley, Jones, & Navaratnam, 1990).

Band 3 represents a terminal mathematics course, some subjects of which are not accredited or publicly examined in the way of the previous courses; these subjects tend not to be associated with high tertiary entrance scores. In 1990, 31 percent of students were in this band of mathematics education (Ainley, Jones, & Navaratnam, 1990;).

As noted earlier, there is considerable variation between states in terms of participation rates as well as type of mathematics studied, so generalizations about mathematics participation on the basis of national statistics are not possible. There are, however, several discernible trends.

One interesting effect observable in recent years has been the decline in number of students undertaking the first band, advanced mathematics, whereas more students are enrolling in the second, more general, mathematics course. Historically boys were much more likely than girls to enrol in Band 1 subjects, while girls enrolled in numbers similar to boys in Band 2 subjects. The predictive validity of school mathematics for success in tertiary mathematics is demonstrably connected to enrolments in the Band 1 advanced mathematics stream. It is standard practice in Australian universities for first-year university mathematics to assume knowledge of the Band 1 subject matter and to teach from that position. While it is technically possible for a student to enrol in higher mathematics without having done the double mathematics option in Year 12, such students have significantly less likelihood of success than those who have taken the Band 1 course.

GENDER AND PARTICIPATION IN SENIOR SCHOOL MATHEMATICS

All of the above features have implications for the question of gender and mathematics education. In Australia, as in other parts of the English-speaking world, there has been considerable attention drawn to the issue of equity in educational outcomes generally and in mathematics education in particular. The situation of girls' under-representation in the senior years mathematics classes is repeatedly given as evidence of continuing inequity in schooling outcomes. Early articles on the subject in this country, as elsewhere, were characterized as exercises in "blaming the victim." In this approach, girls were blamed for their lack of sense and foresight in making inappropriate choices in their senior school curriculum. If only they would see the rightness of studying mathematics all would be well. In this vein programs were initiated in several states designed to encourage girls and young women to "stay" in mathematics classes — "Mathematics increases your options" is one popular slogan.

However all this advertising material is less than specific about which mathematics courses are being encouraged and which options are being increased. The goals of such programs have not been clearly set beyond the national target of getting the same number of girls in mathematics classes as boys. Mathematics educators have yet to address the question of whether their brief is to encourage and facilitate mathematics education across the board for all students or to provide the stimulus for those students with great mathematical potential to pursue high levels of educational achievement. Of course the response could well be that it shouldn't be a question of either/or, but the structures, as outlined in the earlier paragraphs, constitute mathematics education in Australia as falling into one or other camp. If our goal is to get more women to study tertiary mathematics or mathematics-related courses then clearly the aim should be to get more girls into the double mathematics option in senior school. In the Australian context, senior school mathematics still operates as critical filter which determines post-school options. And, as several writers have pointed out, the filter operates with a pronounced gender effect (Gill, 1984; Sells, 1980). However if the aim is to add to the mathematics knowledge of the general populace then the more general mathematics options are the more appropriate target. Until the situation is clearly addressed, it is our suspicion that mathematics education will continue to serve as a critical selection device embedded within the senior school curriculum.

The three bands identified above represent different populations of senior school students and are associated with different gender mixes. Traditionally the second and third band contained higher proportions of female students, whereas the top band routinely enrolled more than three boys for every two girls. As noted by Ainley and colleagues:

> Nearly one in four Year 12 males took advanced mathematics compared to just under one in eight Year 12 females. In contrast females tended to have a higher participation rate in fundamental mathematics programs ... the trend is clear. Males tend to be enrolled to a greater extent than females in advanced levels of mathematics. (Ainley, Jones, & Navaratnam, 1990, p. 28)

And of course the under-representation of girls in senior school advanced mathematics means, in the Australian context, that females will be under-represented in university mathematics, science, and engineering courses and in the ranks of professional mathematicians.

The recent changes in the composition of the senior school years outlined above have been associated with a more even gender balance in the second and third band, no doubt in part as a result of vigorous campaigns to encourage more girls to remain in mathematics education in the senior school. The top band continues to be male dominated in roughly the same proportions across the states, a feature which remains despite the differences in content of the mathematics courses and despite the encouragement of teachers and the wider community to support girls who wish to continue in mathematics. There has been some speculation that attendance at single-sex schools is associated with girls' staying in academic mathematics. This speculation has not been borne out by the research findings (ESSA, 1993; Gill, 1988). One Victorian study revealed that girls decided against continuing with this band of mathematics in equal proportions across three different socio-economic groups — in other words advanced mathematics was no more popular with girls from the higher socio-economic backgrounds than with girls who were less well off (Yates & Firkin, 1986). More recently, a finding from Ainley, Jones, and Navaratnam's work conflicts with this result and shows that, whereas for boys participation in mathematics subject is significantly linked to prior achievement in mathematics and not significantly linked to social background, for girls there is an interactive effect such that girls with high socio-economic status participate in mathematics according to their achievement in mathematics, while girls with low socio-economic status choose against participation in mathematics whether or not they have been successful at mathematics in the past (Ainley, Jones, & Navaratnam, 1990). This issue of girls' experiences and enrolment in math-

ematics would appear to be an area that warrants more research if programs
to encourage girls to enrol in advanced mathematics are to be successful.

The situation relating to gender and participation in mathematics in the
senior secondary years is shown in Table 1 below.

Table 1: Mathematics participation in the final year of
school across the states in 1991

State	Band	Male N	% of cohort	Female N	% of cohort
Q*	1	3641	11	1655	5
	2	8876	27	7117	22
	3	7774	23	9983	30
WA	1	1409	7	660	3
	2	4460	22	4585	23
	3	1812	9	2214	11
SA	1	1526	8	737	4
	2	2242	11	2135	11
	3	2602	14	2705	14
NSW	1	2743	5	1463	3
	1a**	5408	10	4671	8
	2	8781	15	10316	18
	3	8301	15	11039	19
NT	1	74	6	36	3
	2	163	13	128	10
	3	145	11	111	10
Tas.	1	192	5	82	2
	2	668	17	518	13
	3	707	18	780	20
Vic.*	1	5569	11	3186	6
	2	11048	22	11117	22
	3	17224	35	10658	21
ACT	1	788	18	758	18
	2	793	19	927	22
	3	373	9	348	8

*Note: In some states, students may enrol in more than one form of tertiary entry math-
ematics (e.g., Queensland and Victoria); more generally, however, students must
choose between bands.

**In NSW there are four bands with the inclusion of inter-advanced mathematics.

(All statistics obtained from ESSA study and in consultation with state authorities)

In each state, the academic mathematics option is represented as Band 1,
the accredited mathematics for tertiary entry but less advanced level is Band

2, and the more general mathematics that is not accredited for tertiary entry is listed as Band 3. The figures given are taken from a larger report and apply to 1991 (ESSA, 1993); the pattern that they represent has been seen to be consistent for the previous five years at least, given the changes previously noted which have involved fewer enrolments in the advanced band (despite a general increase in the size of the student cohort) and more enrolments in Band 2. The 1991 figures were collected as part of a nationwide study of participation and achievement in senior school and represent the first account of national patterns of student behavior at this level. Figures from several individual states since 1991 show that the situation has not altered substantially.

From Table 1 it can be seen that there are clear imbalances between male and female enrolments in the advanced mathematics band and that males tend to enrol in this band roughly twice as frequently as females in all states except the Australian Capital Territory. Equally evident is the fact that this band attracts a smaller proportion of senior mathematics enrolments than the more general Band 2.

A disturbingly large proportion of senior school leavers in Australia take no mathematics as part of their final year of schooling. In their report on the 1989/90 cohort, Ainley et al put this figure at 17 percent nationally but there were huge variations between states, with 39 percent of Victorian students and 56 percent of Tasmanian students taking no mathematics in their final year of school (Ainley, Jones, & Navaratnam, 1990, p. 88). By 1991 the structure of the senior mathematics curriculum underwent some change in an effort to raise the number of students taking some form of mathematics. While it is pleasing to see more students taking mathematics courses, the restructuring has not affected the gender imbalance of the advanced mathematics courses.

GENDER AND ACHIEVEMENT IN SENIOR SCHOOL MATHEMATICS

There is some research evidence to support the theory that senior school mathematics is the choice of the more able students, thereby justifying the ubiquitous term "hard" frequently used to describe the advanced mathematics subjects. The national study of achievement levels in a range of senior school subjects showed that enrolment in the advanced mathematics band was associated with superior achievement in the form of high tertiary entry score (ESSA, 1993). This effect was particularly pronounced in the case of female enrolments, suggesting that the girls who undertook advanced mathematics in their final year of schooling were a more select group than the boys who followed this course. The second band of mathematics enrolments was also linked, albeit less strongly, to high achievement as seen in a high tertiary entry score, and this effect was also much stronger for females than for males (ESSA, 1993). This result gives rise to the suggestion that girls who are capable of the more advanced mathematics are not choosing to enrol in the subject in the same proportions as their male peers. The girls who undertake higher level mathematics would appear to be a more select group than boys who do so, and the higher overall success rate of the girls is perhaps not surprising. The gender differences in patterns of performance has been summarized as follows:

> Reasonably consistent across all mathematics courses is that there are fewer females at the bottom end of the achievement range than there are males. Female scores tend to be in the middle and towards the top of the range. Males tend to be at the two extremes, with higher proportions performing very well or very poorly. (ESSA, 1993, p. 122)

The not-unfamiliar finding that girls as a group do not appear as often as boys as the top achievers continues to concern feminist mathematics educators. In the outcry following Benbow and Stanley's widely publicized findings with respect to the under-representation of girls and women in the highest levels of mathematical achievement, there was a renewed interest in the ways in which mathematics functions as a male domain. It still remains to be shown *how* the image of mathematics as a male domain constrains the performance of highly able girls and women who are studying the subject. Is mathematics pedagogy the cause of the different levels of achievement? Are the assessment practices in mathematics unwittingly favoring male students? Such questions call for further investigation of mathematics classrooms and related assessment practices.

GENDER AND MATHEMATICS ENROLMENTS: SOME EXPLANATIONS

The recurring gender difference in enrolment patterns in higher level school mathematics presents a more straightforward question. Research has addressed this issue from a range of theoretical locations. Early research in this area looked at school timetables and teacher conceptions of desirable students. Attention was directed to counselling practices operating within schools. Anecdotal evidence implied that the counsellors were frequently instrumental in directing female students away from advanced mathematics and, at the same time, directing male students towards this subject area. Leder's early work on motivation theory offers another explanation for the observed differences between male and female students in deciding to continue with higher mathematics. This branch of motivation theory proposed that academic success for female students was potentially threatening to their femininity, a feature that was replicated in studies of Australian students (Gill, 1980; Leder, 1980). Subsequent work on locus of control offers an account of the different self-perceptions of male and female students and relatedly their difference in confidence levels. Single-sex classrooms and schools have been advocated as providing more effective and successful mathematics education for girls, a position not borne out by careful research on this topic (as noted earlier). Explanations in terms of self-esteem were taken up vigorously in the mid-eighties; however, such approaches tended to be associated with popular conviction and associated commitment on the part of teachers and are not necessarily sustained by rigorous research. These approaches failed to establish why, for instance, the persistent differences in self-esteem had more effect on mathematics enrolment than for example on English or any other senior subject. Relatedly the concept of mathematics as a male domain has been taken up by feminist educators, although the meaning of the concept is less than clear. Some feminist educators have challenged the importance of mathematics as a field of cognitive endeavor and sought to celebrate the intellectual worth and merit of traditional areas of female achievement such as literature and history. While prepared to acknowledge the constructed nature of intellectual value, we feel it would be most unfortunate if this notion occasioned widespread turning away from mathematics by girls and women in education, rather than attempting to reconstitute the subject in ways more inclusive of female interests and experience.

SUMMARY OF THE SCHOOL SITUATION

In summary, the situation regarding gender and mathematics education in Australian schools is very similar to that in other parts of the English-speaking world. There has been a consistent trend for more males than females to enrol in advanced mathematics courses in their senior school years. In Australia, because of the lock-step connection between school subjects and tertiary studies, this situation means that girls and women will continue to be under-represented in tertiary mathematics courses and related fields of employment. The issue raises questions about the goal of mathematics education generally, as well as directing interest towards the ways in which girls and women are positioned vis-à-vis this particular form of high-status knowledge. Gender differences in attitudes to and achievements in mathematics education have been consistently found in education research in the same pattern as the gender differences in participation in senior mathematics courses. Several theories have been proposed to explain these differences, but they continue to occur in the face of programs explicitly designed to counteract their effects. It seems particularly important to try to establish the degree to which actual school experience contributes to the perpetuation of gender differences in mathematics learning. More research is needed if we are to account for the ways in which these differences are produced in terms of both classroom experience and broader understandings of valued knowledge.

HIGHER EDUCATION

Post-secondary education in Australia is organized around two sectors, the higher education sector, consisting of the universities, and a non-degree granting vocational and training sector. Each system has a significant number of publicly administered institutions, and this feature of post-secondary education makes the examination of the education policies and initiatives of government particularly important. The higher education sector, in particular, is almost entirely a system of public institutions as only 2 of the 38 universities currently within it are private; furthermore, the private universities are of recent establishment, offer courses across a narrow range of disciplines, and have relatively low student enrolments. The classification of post-secondary education that does not take place at a university as a single sector is perhaps to suggest a structure more formally defined than is currently the case; however, there does exist within this sector a very large public system of over 200 colleges of technical and further education (the TAFE system) which is currently undergoing rapid growth in terms of student numbers and function. Future discussions of mathematics education will need to take into account developments within this sector.

The present single system of universities is of recent origin and is one of the outcomes of major policy initiatives undertaken by the federal government during the 1980s. The higher education sector between 1965 and 1987 existed as a binary system of universities and colleges of advanced education. The colleges of education were originally conceived of primarily as teaching institutions providing tertiary education with a strong vocational orientation to the level of diploma. However, with the passage of time, this sharp distinction of function became blurred; some colleges became degree-granting and had significant research programs, and in line with international developments, the binary system was formally abolished. A detailed account of the development and subsequent demise of the binary system is given in the report on higher education by the federal government's Department of Employment, Education and Training (DEET), *National Report on Australia's Higher Education Sector* (1993a).

The abolition of the binary system was not simply a bureaucratic exercise in redesignating the existing colleges as universities but usually involved mergers between colleges to create the new universities and, in some cases, involved colleges merging with existing universities. As the federal government used the opportunity that the reorganization of the system provided to introduce significant changes to the relationship between government and the higher education sector — including increased accountability and new funding arrangements — all university administrations in the recent past have expended considerable effort in dealing with the new system. In the short term, the process of adjustment to the new system has introduced a further complication to the implementation of earlier reforms of the system, including the various initiatives undertaken to open up the system to women, and may have affected the rate of change in these areas.

Women in Higher Education

Although the argument is well founded that the organization of the institutions of higher learning into a binary system was in many respects artificial, and also unjust as it entailed differences in funding arrangements, it nevertheless is the case that the binary system existed for a considerable period of time during which the institutions within the two groups evolved into environments that were significantly different and about which useful generalizations could be made. For this reason, a classification of current universities into three groups — largely depending on whether the university had been incorporated before, during, or after the binary system — has been adopted for the presentation of statistics in this discussion. The ten universities in the first group include four which were established before federation in 1901. The universities of this group offer a wide range of courses, have a strong research focus, significant proportions of their students undertake research-based higher degrees, and partly because of their long-standing traditions, are the most prestigious. The second group, the universities established during the period of the binary system, consists of nine universities. These universities, some of which are based in metropolitan areas, while others are based in large regional centres, do not in general enjoy the same prestige as those of the first group or, from a different perspective, are regarded as less elitist. The post-binary system of universities of the third group, 19 in number in 1994, are a diverse group of institutions. As most of these have been formed from former colleges of advanced education, the present characteristics of these institutions still largely derive from their earlier history. The colleges of advanced education were either teachers' colleges or institutes of technology offering specialist courses with a strong vocational orientation — usually in engineering, applied science, and business — and for much of their history, the vast majority of their staff did little if any research.

The universities within the Australian system are in many respects different in character, and two statistics given in Table 2 serve to underline this point. Evidence for the previous observation that the oldest universities of group 1 tend to be the most research oriented is provided by the figures for the proportion of students enrolled in research-based higher degrees; the average proportion of students in this category for group 1 was 8.2 percent, in contrast to the newest group of universities where the proportion was 2.4 percent. Also, the figures in Table 2 show that the oldest universities tend to attract the students who perform best at school; the proportion of students entering these institutions with a tertiary entrance (TE) score in the top quartile of the state in which the university is based was 55.6 percent. There are many other statistics which could be given to show the various ways in which enrolment patterns at universities within the system differ. In the context of our discussion, just as doing the "right" mathematics courses at high school largely determines the opportunities for further study in courses requiring mathematics, studying at a research-oriented university is likely

Table 2: University students

	Total student enrolments (1)	Percent who are female (1)	Percent in research higher degrees (1)	Percent with TE scores in state top quartile (2)
Universities established 1851–1959				
range	10,200–36,500	45.6–56.2	4.6–9.7	31.6–86.2
group average	21,600	50.7	8.2	55.9
Universities established 1960–1986				
range	8,000–24,400	46.3–63.2	2.1–7.5	0.2–16.5
group average	14,300	57.0	5.1	6.9
Universities established 1987–1992				
range	3,800–24,700	46.3–75.2	0.2–4.7	0.1–25.2
group average	14,300	53.4	2.4	6.8
Australia	575,600	53.4	4.9	n.a.

These data refer to the system of universities as it was in 1993. The University of New England, Bond University, and Notre Dame University are not included because of data deficiencies.

(1) These data refer to all students enrolled in 1993. A "research higher degree" has a research project and thesis as 50 percent or more of the course workload.

(2) These data refer to students commencing university in 1992. High school students seeking university entry receive a tertiary entrance (TE) score b⌐s⌐d on their academic performance at school.

(Source: DEET (1994) *Diversity and Performance of Australian Universities, Higher Education Series, Report No. 22*)

to be important for female students wishing to keep open the possibility of a career as an academic mathematician.

The figures presented in Table 2 show that females constituted 53.4 percent of the student population of just under 600,000 students at Australian universities in 1993; also, although considerable differences did exist between institutions with the proportion of students who are female varying between 46.3 percent and 75.2 percent, female students were in the majority at many universities.

While Table 3 clearly shows that female and male enrolments for the various disciplines are skewed, the fact that the female presence is significant at the institutional level has meant that not only is it accepted that women students participate in student affairs at every level but also that there are many women available to do so. The differences between institutions with respect to female enrolment mainly reflect differences in the range of courses the institutions offer. As can be seen in Table 3, female students dominated areas such as education, health, and the arts, where they accounted for about 70 percent of the students, and were under-represented in science, engineering, agriculture, and architecture. As the broad field of science in Table 3 included mathematics, the proportion of females enrolled in mathematics was somewhat lower than 38.8 percent but not as low as in engineering where it was 10 percent. Also worthy of note is that in every field, the representation of women was significantly lower at the postgraduate level. While these figures indicate progress — for example, in 1979 the proportion of science students who were female was 32.4 percent — both the general features of the enrolment patterns of female students observed in many countries — underrepresentation in non-traditional areas, including science and engineering, and the phenomenon of "cascading losses," where women are proportionally fewer at each level and stage of education — are also characteristics of the enrolment patterns of female students at Australian universities.

A comprehensive analysis of enrolment statistics and a survey of Australian research in this area is given in Powles (1987).

Table 3: Females' share of total enrolment
by field of study and level, 1990

Field	Percent undergraduate	Percent postgraduate	Total
Agriculture	33.9	27.7	33.0
Architecture	34.9	26.8	33.4
Arts	69.0	60.9	68.0
Business	42.0	32.2	40.7
Education	75.0	66.4	72.4
Engineering	10.2	9.6	10.1
Health	73.6	62.4	72.2
Law	46.3	40.8	45.2
Science	40.2	31.2	38.8
Veterinary Science	55.8	35.9	52.5
Non-award	n.a.	n.a.	53.3
Total	53.6	48.0	52.7

Courses at Australian universities are classified into ten major fields.
(Source: DEET, *Female Students, Higher Education Series*, Update No. 1, 1991.)

Table 4: University academic staff

	Total academic staff (1)	Percent who are female (1)	Percent above senior lecturer who are female (2)	Quality assessment band (3)
Universities established 1851–1959				
range	780–2330	21.1–34.7	7.7–21.3	1–3
group average	1590	28.3	11.6	1.4
Universities established 1960–1986				
range	420–1330	27.4–37.9	12.0–23.9	2–5
group average	760	33.9	16.4	3.8
Universities established 1987–1992				
range	510–2580	26.7–56.9	8.3–37.6	3–6
group average	1400	35.1	18.7	5.0
Australia	32,210	32.6	15.4	

These data refer to the system of universities as it was in 1993. The University of New England, Bond University, and Notre Dame University are not included because of data deficiencies.

(1) These data refer to 1993 staff.

(2) These data refer to 1992 staff.

(3) Institutions ranked on scale of 1 to 6 by 1993 quality audit.

(Source: DEET (1994) *Diversity and Performance of Australian Universities, Higher Education Series, Report No. 22.*)

The figures presented in Table 4 show that in 1993, 32.6 percent of academic staff at Australian universities were women. Highly aggregated figures, however, only give part of the picture, and as the figure of 15.4 percent for the proportion of staff classified as above senior lecturer and female shows, women were concentrated in lower-ranked positions. Women were also concentrated in lower-ranked universities. Table 4 shows that in the 1993 quality audit, the newest universities in general received the lowest rankings and that these institutions typically employed high proportions of female academic staff. Other statistics which show the extent of the disadvantage that women academics continue to suffer in employment include the over-representation in non-tenurable and part-time positions (see DEET, 1993b).

In the context of the present discussion, it would have been relevant to have commented on whether the statistical patterns of employment of women academics across the various disciplines suggest that women academics in mathematics departments and other non-traditional areas are especially disadvantaged; however, the statistics disaggregated to this level are not particularly reliable and are only now becoming easily accessible. It is perhaps not unrelated that the professional society for Australian mathematicians, the Australian Mathematical Society (AMS), has not yet directly addressed the issue of women in mathematics. Although this absence of official concern for women students and academics by the AMS is somewhat dismaying, the fact that an eminent women mathematician has just completed a term of office as president of the society shows that it is, nevertheless, possible for a women mathematician in Australia to succeed in academe and be accorded recognition by her peers for her achievements.

EQUITY AND ACCESS INITIATIVES

The women's movement of the 1970s did much to raise consciousness about the position of women in Australian society, and following a period of intense public debate, the federal government and the state governments between 1977 and 1987 passed legislation concerning sex discrimination, affirmative action, and/or equal opportunities. The fact that principles of equality have been enshrined in legislation for over a decade now, in many instances, has been important in raising the status of Australian women in general. The enactment of the legislation also forced universities to deal with women's issues at the institutional level as they were required, for example, to appoint equal opportunity officers. The pressure for universities to address these issues came from within as well, and during this period of public debate and legislative activity, universities were conducting their own inquires. Women academics led and contributed to many of the debates and given the situation of academic women in the 1970s so well analysed in the book *Why so few? Women Academics in Australia* (Cass et al., 1983), it is hardly surprising they agitated for change. In the final chapter of this book, the authors made a number of recommendations for improving the position of women academics in universities. It is telling that these recommendations — which included providing for fractional full-time appointments with conditions equivalent to full-time tenured staff, providing for paid and unpaid parental leave, providing for childcare facilities, reforming discriminatory conditions in superannuation schemes, establishing systems to monitor and report on equal opportunities, and introducing women's studies courses — now seem obvious and have been largely implemented.

The climate of opinion towards equal opportunities issues is such that all universities now describe themselves as equal opportunity employers. And from the point of view that they have all incorporated equal opportunity principles into their formal procedures for appointing and promoting staff, perhaps they are. Women academics, however, are not progressing through the system at the same rate as their male counterparts, and universities may have a further responsibility to introduce measures at the institutional level to deal with the other factors impeding their progress. Indeed, some universities are currently involved with projects which target female staff. While most of these projects adopt a staff-development approach, some have involved introducing formal mentoring schemes. There has been particular interest in the role mentors play in higher education in Australia, following in part the work of Byrne (see Byrne, 1993, for a detailed discussion of mentoring processes and references to earlier studies).

Universities are currently using a variety of strategies to increase the number of women in courses where they are significantly under-represented — for example, science, engineering, and technology, and higher degrees, particularly research. The low participation rates of women in these areas are seen as an equity issue by the federal government, and in the government's plan for equity in higher education, *A Fair Chance for All* (DEET, 1990), women in non-traditional areas constitute one of the groups identified as experiencing significant disadvantage in access to high education. The national plan identifies objectives and targets for groups designated as disadvantaged and discusses strategies considered either to have been successful in the past or likely to be successful in the future in addressing their needs. The strategies suggested for achieving the objectives and targets for women identified in the plan include providing information to girls about non-traditional courses and careers, establishing links between universities and schools in their catchment areas, providing bridging courses in mathematics and science, reviewing curricula and teaching processes, developing academic and counselling support programs, establishing mentoring schemes, providing scholarships to undertake postgraduate studies part time, providing postgraduate scholarships for women, and establishing women's networking groups.

As all publicly funded universities are required by the federal government to provide equity plans as part of the educational profile they submit for funding purposes, the government has been able to produce an information paper to compare, among other things, the strategies the various universities were planning to use in the triennium 1992 to 1994 (DEET, 1993c). The information paper is superficial; however, it does provide an overview of Australian programs during that period. The paper found that 23 universities were providing bridging programs, 32 were involved in awareness programs, and 23 had initiated support programs for women in non-traditional areas. (Although the total number of planned programs was large, many of the programs were reasonably modest.)

As it would be beyond the scope of this discussion to describe all the projects the universities put in place for their equity programs, a few projects which have been developed partly to overcome poor backgrounds in mathematics will be briefly described. The New Opportunities in Tertiary Education program at Queensland University of Technology is a comprehensive bridging program, offered part time over a full year, which includes a component of remedial mathematics. (Since many boys also fail to take appropriate mathematics courses at school, universities commonly offer bridging programs.) Some universities offer programs which, although not exclusively reserved for women, provide them with valuable support in mathematics. The numeracy skills learning programs offered at Macquarie University is one example, and the Mathematics Learning Centre established at The University of Sydney is another. It is interesting that at Macquarie two-thirds of the students enrolling in courses such as the Maths Assertive Course are women, and at Sydney two-thirds of the students using the Centre's services are women in science-related courses.

The question arises as to what has actually been achieved by all of this activity. Progress is hard to judge because statistics on female participation rates by institution, disciplines area, and level are not routinely available. From the statistics that are available, however, it is clear that there are significant inter-institutional and interdisciplinary differences in female participation rates which have yet to be explained (Byrne, 1993; DEET, 1993c). The question also arises as to whether other projects might have been more effective in promoting change. While the federal government is to be commended for recommending equity programs, the approaches they have suggested have not always been informed by research. It is argued that projects which are underpinned by research and deal with a group of related factors rather than one factor at a time are likely to be more effective (Byrne, 1993).

SUMMARY OF THE SITUATION IN HIGHER EDUCATION

The position of women academics in Australia does not differ significantly from that which is observed in other parts of the English-speaking world. Although their situation has improved since the enactment of federal and state legislation concerning sex discrimination, affirmative action and/or equal opportunities, and other initiatives taken by the universities, women academics remain concentrated in lower-ranked positions at lower-ranked universities. Whether women academics in mathematics departments appear to be especially disadvantaged cannot be determined from the available data. The enrolment patterns of women in university courses also are similar to enrolment patterns observed elsewhere. Women are under-represented in non-traditional areas including science (mathematics is included in science in Australia University data) and engineering. The phenomenon of "cascading losses," where women are proportionally fewer at each level and stage of education, also applies. Various programs have been introduced by universities to enhance the access women to non-traditional courses. As a lack of an appropriate background in mathematics is often a problem, universities have responded by developing programs to deal with this deficiency. The federal government has been particularly active in promoting equity in access projects. These projects although well-intended may not be particularly effective.

REFERENCES

Ainley, J., Jones, W., & Navaratnam, K. K.(1990). *Subject choice in senior secondary school.* ACER. Department of Employment, Education and Training. Canberra: AGPS.

Byrne, E. M. (1993). *Women and science: The snark syndrome.* London: Falmer Press.

Cass, B., et al. (1983). *Why so few? Women academic in Australian universities.* Sydney: Sydney University Press.

Department of Employment, Education and Training. (1990). *A fair chance for all: higher education that's within everyone's reach.* Canberra: AGPS.

Department of Employment, Education and Training (1993a). *National report on Australia's higher education sector.* Canberra: AGPS.

Department of Employment, Education and Training. (1993b). *Female academics* (Higher Education Series Report No. 18). Canberra: Author.

Department of Employment, Education and Training. (1993c). *Institutional equity plans 1992–1994 Triennium.* Canberra: Author.

ESSA. (1993). *Final report of the gender equity in senior secondary school assessment project. A project of national significance.* Canberra: DEET

Gill, J. (1980). *Fear of success: A theoretical review and an empirical study in an Australian high school setting.* Unpublished master's thesis, Adelaide University.

Gill, J. (1984). *Hanging Rock revisited: The not-so-mysterious disappearance of girls along certain educational paths.* In R. Burns & B. Sheehan (Eds.), *Women and education.* ANZCIES Proceedings, Latrobe University, Vic.

Gill, J. (1988). *Which way to school? A review of the evidence on the single sex versus coeducation debate and an annotated bibliography of the research.* Canberra: Commonwealth Schools Commission.

Jones, W. (1986). *Secondary school mathematics and technological careers* (ACER Research Monograph No. 32). Hawthorn, Vic.: ACER

Leder, G. (1980). Bright girls, mathematics and fear of success. *Educational Studies in Mathematics, 11,* 411–422.

Powles, M. (1987). *Women's participation in tertiary education: a review of recent research.* (2 ed.). Melbourne: University of Melbourne Centre for the Study of Higher Education.

Rosier, M. J. (1980). *Changes in secondary school mathematics in Australia 1964–1978* (ACER Research Monograph No. 8). Hawthorn, Vic.: ACER.

Sells, L. (1980). The mathematics filter and the education of women and minorities. In L. H. Fox, D. Tobin, & L. Brody (Eds.), *Women and the mathematical mystique.*

Baltimore: Johns Hopkins University Press.

Yates, J., & Firkin, J. (1986). *Student participation in mathematics: Gender differences in the last decade*. Melbourne: VCAB.

MEGAN CLARK

MATHEMATICS, WOMEN, AND
EDUCATION IN NEW ZEALAND

New Zealand is a country in the South Pacific with a small population (3.5 million) in an area about the size of Great Britain. It consists of two main islands and many smaller ones so this population is quite widely dispersed. This dispersion has consequences for education. For example, New Zealand has seven universities with student populations of approximately 10,000, apart from the University of Auckland which has 20,000. The indigenous Maori population was colonized by Great Britain in 1841, and this colonial legacy has left its imprint in our education system which is, in some ways, similar to that of Scotland.

About half of New Zealand's young children receive some form of early childhood (preschool) education, which is available from a range of providers. In the main these are free kindergartens, play centres (run by parent collectives), fee-paying childcare centres, and "language nests" providing immersion in Maori (or other Pacific Island languages).

Education is compulsory from the ages of 6 to 16 (15, prior to 1993) but most children start school on their fifth birthday. Most schools are state schools, some of which have a religious nature. Virtually all public primary (elementary) schools are co-educational, and grouping students by ability is common in most classrooms. Primary school lasts for eight years, followed by secondary (high) school which may last a further five years. Secondary schools may be co-educational or single sex. Some use traditional streaming of classes according to ability, while others use mixed ability classes. Recently some schools have instituted bilingual units or immersion teaching in Maori for some classes. This development is expected to proliferate. Rural pupils may be enrolled in the Correspondence School (as may prison inmates and hospital patients). Where possible, disabled students are mainstreamed into ordinary classrooms. Education is free up until age 19. After secondary school, about 45 percent of school leavers continue with formal tertiary education, which may be at a polytechnic (about 20%), a college of education (2%), or at a university (23%). Colleges of education offer teacher training, while polytechnics offer a range of vocational programs. There has been a large increase (approximately 37%) in the total number of students enrolled in tertiary education over the last five years, reflecting changes in the type and number of jobs available and a greater expectation that students might continue with education after secondary schooling. Tertiary education is partially subsidized with students currently paying about 15 percent of fees.

G. Hanna (ed.), Towards Gender Equity in Mathematics Education, 257–270.

In 1991, government spending on education as a proportion of Gross Domestic Product was 6.1 percent, the same as the Organisation of Economic Co-operation and Development (OECD) country average. Of this expenditure, about 5 percent is on the preschool sector, 50 percent on schools, 35 percent on tertiary education, and 10 percent on other education.

POLICY

The new school mathematics curriculum, which took effect in 1993, is quite explicit in its policy on women students:

> It is a principle of the *New Zealand Curriculum Framework* that all students should be enabled to achieve personal standards of excellence and that all students have a right to the opportunity to achieve to the maximum of their potential. It is axiomatic in this curriculum statement that mathematics is for all students, regardless of ability, background, gender, or ethnicity....
>
> In many cases in the past, students have failed to reach their potential because they have not seen the applicability of mathematics to their lives and because they were not encouraged to connect new mathematical concepts and skills to experiences, knowledge, and skills which they already had. This has been particularly true for many girls, and for many Maori students, for whom the contexts in which mathematics was presented were irrelevant and inappropriate. These students have developed deeply entrenched negative attitudes towards mathematics as a result....
>
> The suggested learning experiences in this document include strategies that utilise the strengths and interests that girls bring to mathematics. Techniques that help to involve girls actively in the subject include setting mathematics in relevant social contexts, assigning co-operative learning tasks, and providing opportunities for extended investigations.
>
> The suggestions also describe experiences which will help girls develop greater confidence in their mathematical ability. Girls' early success in routine mathematical operations needs to be accompanied by experiences which will help them develop confidence in the skills that are essential in other areas of mathematics. Girls need to be encouraged to participate in mathematical activities involving, for example, estimation, construction, and problems where there are any number of methods and where there is no obvious "right answer." (Ministry of Education, 1992)

THE MATHEMATICS CURRICULUM

The curriculum in mathematics is prescribed at the national level by the Ministry of Education. It is prescribed for public schools at the primary and secondary levels and for private schools at the secondary level. The curriculum has several general goals, such as the desirability of positive attitudes to mathematics and of becoming a "problem-solver" and is more process-oriented than previous curricula. In the past, junior, middle, and high school had separate curricula. The current curriculum is much more holistic than previous ones, even in the fact that it extends from new entrants in the primary school to the end of Year 13. It is organized into six strands but this does "not mean that the mathematics is expected to be learned in 'discrete' packages" (Ministry of Education, 1992). Students are expected to inter-weave the strands and make connections. These six strands are: *number, geometry, measurement, algebra, statistics*, and *mathematical processes* (reasoning, problem solving) and, in the senior secondary school, calculus replaces number. These strands are intended to be taught separately part of the time and integrated otherwise. Mathematical processes, in particular, are to be applied across all content areas.

The curriculum defines learning outcomes in eight progressive levels of achievement. These learning outcomes are specified for each pair of school grades and, in theory, are independent of age. Attempts have been made to include ethnomathematical material in the current curriculum, and various strategies are used to encourage the participation of girls such as an empha-sis on co-operative rather than competitive learning.

Monitoring of national educational standards in mathematics is due to begin in the near future. It is proposed that samples of children at ages eight and 12 be assessed and that this assessment be used as a guide to the state of the system as a whole.

PARTICIPATION AND PERFORMANCE

Primary (Elementary) School: Years 1 to 8

Mathematics is compulsory at this level for all pupils. In terms of perfor-mance in mathematics, girls do better in computational areas and boys in word problems, but these differences are small. Young-Loveridge (1991), tracking children's number concepts through the first four years of school, found minimal differences between high-achieving boys and girls but in the low-achieving group, boys were much more likely to overcome early diffi-culties than girls.

Secondary (High) School: Years 9 to 13

At the beginning of secondary school, mathematics is compulsory for all students. For some three decades New Zealand suffered from a severe shortage of qualified (i.e., mathematics majors) mathematics teachers. The situation is now improving (Clark, 1993a) but this is still a problem for 10 to 20 percent of mathematics classes at secondary school. Currently mathe-matics teachers are just as likely to be female as male, whereas 15 years ago two-thirds of them were male. Few comparative studies of mathematics achievement have been made, but the 1981 IEA study was repeated on a large sample of Year 9 students in 1989. In 1981 there was a small perfor-mance difference in favor of males in this test (p = 0.05) (Department of Education, 1987). In 1989 there were no significant differences in the test scores of males and females, (Forbes, Blithe, Clark, & Robinson, 1990). This reduction in gender difference is entirely because of an improvement in the girls' mean score. Within subsets of the test used (arithmetic, algebra, geometry, and measurement) there were no significant gender differences in 1989 unlike in 1981 when boys scored markedly higher than girls in geometry. The 1989 survey was the first to provide concrete evidence on how standards in mathematics in New Zealand compared to an earlier era. Contrary to widespread public opinion, such standards (insofar as these tests measure them) were not dropping.

Year 11 (Fifth Form)

The fifth form is the last year of compulsory education for most students. It is in the fifth form that mathematics becomes an optional subject for the first time. Over the last 20 years the percentage of girls studying mathematics at this level has more than doubled and now, like that of boys, is almost 100 percent. In the public examination (School Certificate) usually taken at the end of this year, mathematics has ranked second in popularity (after English) as a subject for all students for the last 20 years. Grade distributions in the mathematics exam are very similar from year to year. There have been consistent small gender differences in favor of boys over a number of years.

Year 12 (Sixth Form)

Until the middle of the 1980s, New Zealand secondary schools had low levels of participation after the years of compulsory education, unlike North American countries. Over the last five to ten years, retention rates in the senior secondary school have increased substantially (Pole, 1993) as seen in Table 1.

Table 1: Retention of year 9 entrants to year 13

1985	1993
17%	44%

This increase in participation as a whole has been accompanied by an increase in the participation of young women in upper secondary school mathematics.

The proportion of young women taking mathematics in the sixth form (Year 12) has climbed from 55 to 76 percent over the period 1974 to 1991 (almost all males in the sixth form continue with mathematics). This increase is due to a number of factors. First there has been increased social pressure to continue with mathematics in order to improve the chances of getting a job. Then there have been deliberate efforts by schools to make mathematics more accessible and relevant to girls. The implementation of the Family Math program and the use of EQUALS material in schools have proved very successful. At the same time, it has been deliberate policy of the Ministry of Education to encourage the participation of young women in mathematics, and the year-long "Girls Can Do Anything" campaign, sponsored by the Ministry of Women's Affairs in the early 1980s, had a very high public profile. However, there are some disturbing features about the participation of young women in Year 12 mathematics. Females who sat the public examination in mathematics in Year 11, and who continued on at school to Year 12, were less likely to take Year 12 mathematics than males, regardless of how successful they had been in Year 11 mathematics. Females obtaining the top grades in Year 11 mathematics from single-sex schools are much more likely not to take Year 12 mathematics (19%) than girls from co-educational schools (7%) (Blithe, Clark, & Forbes, 1993). The proportion of boys with such a grade that did not continue with mathematics (8%) is not affected by the type of school attended. For the other grades, school type has little effect on continuation with mathematics. Differences in performance, where they exist, are small. In Year 12, only in the very highest grade do boys do better than girls (Forbes, Blithe, Clark, & Robinson, 1990). Females dominate the middle grades in a pattern familiar in some European countries, notably Britain (Burton, 1993).

Year 13 (Seventh Form)

The number of males and females continuing on to this last year of secondary school are approximately equal (both about 10,000). But the increased participation of females in mathematics seems to have stalled at Year 12. There are two mathematics courses available in Year 13: *Mathematics with Calculus* and *Mathematics with Statistics*. The patterns of participation in these courses are very different as Table 2 illustrates. Both courses have had increases of 2 to 3 percent in Year 13 female participation since 1990.

Table 2: Participation in year 13 mathematics courses, 1992

Enrolled in	Percentage of Year 13	
	Females	Males
Mathematics with Calculus	34	52
Mathematics with Statistics	54	68
both courses	23	40
neither course	35	20

Sixty-five percent of young women took at least one of these courses; a drop in mathematics participation from Year 12 of about 10 percent (New Zealand Qualifications Authority, private communication). Furthermore, approximately 45 percent of students taking *Mathematics with Statistics* are women (Blithe, Clark, & Forbes, 1993) but they make up only 40 percent of *Mathematics with Calculus* students. However, a much larger proportion of boys than girls take both mathematics papers. In 1992, 63 percent of students taking both courses were males while only 37 percent were female. This pattern has persisted for some time.

Much debate in recent years has centred on whether or not girls are disadvantaged in co-educational schools. Certainly some participation differences have been observed in the past. In 1981, in Year 13, the then *Pure Mathematics* paper was taken by 60 percent girls in co-educational schools but by only 48 percent girls in single-sex schools, while for boys there was no such difference. More recently, the proportion of girls doing the *Mathematics with Statistics* course only has been less in co-educational than in single-sex schools, and this is also the case for boys. There are indications that a higher proportion of girls in co-educational schools than in single-sex schools study mathematics in the seventh form, but the differences between the proportions of boys and girls doing two papers in co-educational schools are greater than in single-sex schools.

From 1979 to 1988 there were small but persistent differences in performance in both papers in favor of boys. This was true for girls from both co-educational and single-sex schools. The extensive study by Forbes, Blithe, Clark, and Robinson (1990) found no significant differences in performance between students from single-sex and co-educational schools. In the last few years, the differences in mean scores for males and females have disappeared. A sample of results from the 1993 *Mathematics with Statistics* examination has shown no difference in mean score between males and females. The pattern in the past (Forbes, Blithe, Clark, & Robinson, 1990) has been for any differences to be more marked among the top-scoring candidates — that is, a tendency for boys to "outperform" girls in the high-scoring category. This outcome features even more strongly in the prestigious Scholarship examinations in these subjects taken by only a small proportion of seventh form students (approximately 12%).

Whether or not candidates are also studying *Mathematics with Calculus* seems to be the major factor affecting *Mathematics with Statistics* performance. When *Mathematics with Statistics* candidates doing one or both mathematics papers are analysed separately, gender differences in mean performance almost disappear. Gender differences in *Mathematics with Statistics* performance can thus largely be accounted for by the amount of mathematics studied. A simplistic strategy then is to encourage girls to take both mathematics papers in the seventh form, but this may not be desirable if it ultimately restricts the subject choices of girls at this level.

The preference that young women have (or are encouraged to have) for the Statistics paper is of considerable interest. In the 1988 *Mathematics with Statistics* examination for Year 13, there was a small but significant gender difference ($p = 0.05$) in mean performance in favor of boys (51.1 vs 47.0). In this course, 20 percent of the final overall mark is obtained from an internally assessed "project" component and 80 percent from a three-hour examination. There is a small but consistent tendency for girls to do better on the internally assessed component than on the examination. Much has been written (Blithe, Clark, & Forbes, 1993) on the improved performance of girls in take-home tasks and essay questions, and it is possible that female students prefer the type of assessment available in the *Mathematics with Statistics* course compared with the three-hour examination in *Mathematics with Calculus*.

Consistently, over a number of studies (Clark, 1993b; Purser & Wily, 1992) and for ages ranging from 13 to 20, young women in New Zealand perform better on problems that are couched in terms of a situation that is familiar to them. Problems in the *Mathematics with Statistics* course are much more likely to fit this description than ones in *Mathematics with Calculus*. This poses a dilemma for secondary school teachers of mathematics. Do they encourage better performance and confidence by posing comfortable problems or should they be encouraging young women not to be such dependent learners? If it seems in a female student's interest to study only one mathematics paper in Year 13, should a teacher try to persuade the student to take *Mathematics with Calculus* (necessary for many university courses) or let her take the more attractive (and somewhat easier) *Mathematics with Statistics*?

University

There has been an improvement similar to that in senior secondary school in the proportion of students continuing on to university (see Table 3).

Table 3: Percentage school-leavers going directly to

Any tertiary education		University	
1981	1991	1981	1991
19	45	11	22

In 1992, in the universities as a whole, women made up 54 percent of all first-year students (Ministry of Education, 1993) while in first-year courses in mathematics departments, the proportion ranged from 26 percent to 40 percent. In third-year mathematics courses, this proportion remains 24 to 40 percent. This range seems to reflect the percentage of women among those students who take both mathematics papers in the seventh form. However, some statistics courses have a much higher proportion of female students, reflecting in part the larger participation of women in the Year 13 *Mathematics with Statistics* course. Across the universities, first-year applied statistics courses have a percentage of women in the class ranging from 40 to 58 percent, and third-year classes (which can have less than 10 students) may have 40 to 80 percent women students.

In the case of university papers in mathematics, in 1991 and 1992, there were no significant overall differences in performance in pure mathematics between males and females for the few courses that have been analysed. In these courses, the amount of mathematics studied does not seem to be having a big impact on how well students perform, in contrast to the effect at secondary school (Blithe, Clark, & Forbes, 1993).

At one university, statistics courses have been monitored for many years. In one first-year course women have a very high participation rate (approximately 60%), regularly outperform men, and, in particular, take a higher proportion of the A grades (Clark, 1993a). This proportion is invariably higher than their proportion in the class. The course is in applied statistics with applications (and problem contexts) drawn from the social and life sciences. There is no simulation in this course, and the females perform significantly better than males in both the timed examination and in project work. However, this difference is more pronounced in the investigative, "take-home" project work. In contrast, women have a low (23–35%) participation rate in the other statistics course at this university but still achieve a percentage of the A grades approximately equal to their participation in the course. This alternative course is more abstract, involves some simulations, the bulk of the applications used come from physics and industry, and there is less internally assessed work.

At the graduate level, there are recent encouraging signs: over the universities as a whole, women made up 36 percent of all doctoral students in 1991 (Sturrock, 1993) but in 1993, 31 percent of all 51 doctoral students in mathematics and statistics were women (Morton, 1993).

Polytechnics

Polytechnic enrolments of women have more than tripled in the last 20 years and by 1991, the proportion of female students in polytechnics was 45 percent (Sturrock, 1993). However, these women are concentrated in a rather narrow and traditional range of courses, such as nursing (nurses' training is concentrated in the polytechnics in New Zealand), secretarial training, and hairdressing. Mathematics, in the polytechnics, is studied only as a necessary part of various qualifications such as New Zealand Certificate in Science (for technicians), New Zealand Certificate in Engineering (for technicians) and various trades requirements. All of these have very low levels of female participation.

ACADEMIC EMPLOYMENT IN UNIVERSITY MATHEMATICS AND/OR STATISTICS DEPARTMENTS

Throughout the university system, women earn on average less than men, reflecting a concentration in relatively junior positions. Mathematics and statistics departments are no exception in this regard.

Of the staff at the level of lecturer or above in the mathematics/statistics areas in New Zealand universities, 10 percent were women (17 in 165) in 1993 and 12 percent (22 in 181) in 1994. At least six of the 22 women in 1994 have permanent *part-time* positions compared with three of the 159 men. Perhaps, more tellingly, of the 21 full professors of mathematics or statistics in New Zealand universities, none are women and of the fifteen associate-professors in these areas none are women. In terms of research activity, 12 of the 22 women are to be found in statistics or operations research, almost half that number are to be found in pure mathematics and also in mathematics education. Only two of these women are in applied mathematics. Almost a quarter of university staff in statistics are women, compared with one-ninth in pure mathematics, and one in 20 in applied mathematics. This dearth of women is most noticeable in academics aged over 40 and it is among young staff that improvements have occurred.

How does this compare with university academic staffing as a whole? In 1992, of 3315 staff of the grade of lecturer or above, 23 percent were women (Educational Statistics, 1993), so women are present in mathematics and statistics at only half their rate in the universities as a whole. Over all disciplines in the universities, 16 of 361 full professors are women and 32 of 374 associate professors are women. This has to be read in conjunction with the large increases in the percentage of women in the various levels of university employment over the last 15 years (see Table 4).

Table 4: Females as percentage of total university staff
by academic rank

	1976	1981	1983	1992
Full professor	1.6	2.6	2.9	4.4
Associate-professor and senior lecturer	7.3	7.2	7.7	13.1
Lecturer	13.4	20.0	20.4	42.2

A survey of university staff in 1984 indicated large differences between men and women academics in certain respects (Wilson, 1986). Of those staff who had had an interruption to their studies, 20 percent of the women had their studies interrupted for marriage or family reasons, while none of the men cited this as a reason for interruption. Women tended to be appointed to their academic jobs at a lower level than men, had a lower publication rate, and held less positions of responsibility within the university. Of those academics who required childcare, the majority of women used help from outside their relationship, while the majority of men used their spouse to care for children.

MAORI WOMEN AND MATHEMATICS

There has been a major improvement in Maori participation in the senior secondary school in the last decade (as shown in Table 5), but this improvement has not been as rapid as that for the population as a whole, and the gap between the percentages of Maori and non-Maori continuing on to the senior secondary school has widened (Forbes & Mako, 1993).

Table 5: Maori retention rates from year 9 to years 11, 12 and 13

	1981	1991
Year 11	63.5	85.3
Year 12	23.5	54.9
Year 13	3.2	17.8

Much of this can be attributed to Maori students' lesser participation and performance in the Year 11 School Certificate examination, which has to be passed to allow access to higher level. The subject choices of Maori students and the scaling system used in the past to rank subjects (according to level of "difficulty") mitigated against Maori success. The Ministry of Education (1992) considers that timed, written tasks are not reliable indicators of Maori performance, but the School Certificate examination is a collection of mainly three-hour written papers. Some recent changes may improve the situation; for example, since 1990 the Maori language paper now has both oral (internally assessed) and written components, whereas before it was entirely written.

For mathematics, there is currently in production a Maori version of the new curriculum. This is not just a straight translation of the English version, but one which uses exemplars familiar in Maori life and suggests teaching strategies that have been found to be successful with Maori pupils. It is the first time that such a curriculum has been attempted and mathematics is the first (and so far, the only) subject in which this is being tried.

As only a small group of Maori students is retained to the senior secondary school, the group that continues with mathematics is smaller still. Maori males have a small but consistent advantage over Maori females in the public examinations in mathematics. As Maori males perform less well than non-Maori of either gender, it can be seen that Maori females are, in terms of traditional measures of performance, in a very poor position. While participation in Year 12 mathematics is now running at about 75 percent for females as a whole, for Maori females it is more like 40 percent. (This figure is similar for Pacific Island females in New Zealand, whereas Asian females participate in mathematics at a very high level). Maori women are under-represented in the universities compared to females as a whole, principally because they are twice as likely to leave secondary school without formal qualifications.

Thus, it is a mistake to consider the position of women in mathematics in New Zealand as if "women" were a homogeneous group. The position for Maori women is very different to that of non-Maori, and this is probably true for other groups of women, such as "poor" women.

THE MATHEMATICS COMMUNITY

The mathematics community in New Zealand is remarkably cohesive, more so than any other academic discipline. The country has a collection of regional mathematics associations which, in the main, consist of enthusiastic secondary school teachers of mathematics together with smaller numbers of primary school teachers and university mathematics teachers. These mathematics associations have regular meetings, run workshops, disseminate resource material, and monitor educational issues relevant to mathematics education. The associations rotate hosting the New Zealand Association of Mathematics Teachers Conference which is held every two years. Other venues for the discussion of issues in mathematics education are the New Zealand Association for Research in Education conference and the annual colloquium of the New Zealand Mathematical Society (university and research mathematicians) which has a "mathematics education day" as has the New Zealand Statistics Association as part of the annual conference.

It is the dedicated members of these associations that provide extra-curricular extension material for gifted students. Since 1988 New Zealand has been represented in the International Mathematics Olympiads and, in the period 1988 to 1993, has won ten medals (silver and bronze), two of which have gone to young women.

Three of the universities in New Zealand have specific centres relating to mathematics education, and papers in mathematics education are also offered in a fourth. This is over and above the teacher-training courses offered in the Colleges of Education.

The reasons for the tight links existing between the sectors, despite the traditional gulf between primary and secondary school teachers, are largely historical and due to the efforts of a few dedicated individuals. These links have served mathematics education well when curriculum change has been mooted and have enabled mathematics teachers to pull together against suggested changes in assessment, particularly against movements that would reduce the subject to a collection of isolated techniques.

WOMEN IN EMPLOYMENT

Women's participation in the paid labor force has tripled since the 1950s.

These days, over half the adult female population is in the paid labor force, compared with about one-fifth in 1951. However there are still proportionately fewer women than men in the labor force (Sturrock, 1993). While there have been increases in the proportion of women in professional/technical and administrative occupations, the bulk of women are still to be found in clerical/service/sales areas. This may be a reflection of the subjects chosen by women in secondary and tertiary education. Average incomes of women are lower than those of men, and women tend to hold lower qualifications than men.

A recent survey showed a number of statistically significant gender differences in the use of mathematical techniques in everyday life. Men used calculators, decimal arithmetic, percentages, and statistics more than women. (Knight, Arnold, Carter, Kelly, & Thornley, 1993). However, the majority of mathematical skills identified as being important in the workplace are (with the exception of problem solving) those aspects of mathematics that females tend to excel in: arithmetic, measurement, statistics, calculators, and the use of algebraic formulae.

ATTITUDES

Throughout primary school, female students have lower expectations than males. Boys are more likely to enjoy mathematics and to consider themselves good at it than girls are (Young-Loveridge, 1991). A survey conducted in the senior secondary school (Rivers, Lynch, & Irving, 1989) in 1988 established that over 75 percent of the students aspired to tertiary education, but young women aimed for polytechnics in greater numbers than males and aimed for teacher-training in very much greater numbers. Female students' career choices were predominantly in teaching, nursing, art/design, and secretarial areas, whereas males had a much broader list of aspirations. According to Sturrock (1993), the major contributing factor to this state of affairs is the socialization by the family, school, and society.

Some inconsistency and confusion are exhibited in attitudes to mathematics (and mathematics education) by the general population and by mathematics educators. Young-Loveridge (1991) reports a negative attitude towards mathematics held by many mothers. The remark "I couldn't do mathematics when I was at school" is commonly heard and seems often to be a matter for pride. Currently, many teachers subscribe to the view that education in mathematics should be "fun," and when it gets to the "hard and serious" stuff (Keitel, 1993), many students get an unpleasant surprise and cease to engage in the subject. Disproportionately such students are female.

Those women who have succeeded in mathematics report that they have done so largely unsupported by other women who still seem to have deeply ambivalent attitudes to mathematics (personal communications).

OVERVIEW

There are no particular areas of concern with regard to the performance of women in mathematics in New Zealand, but participation is an issue, especially in Years 12 and 13 (sixth and seventh forms) at secondary school. Reduced participation in mathematics at tertiary institutions is largely a consequence of the drop off from Years 12 to 13. This is the time period which should be targeted in any effort aimed at improving the proportion of women engaged in university study of mathematics.

For Maori women, participation and performance in mathematics (and education, in general) is an area of real concern, and much needs to be done to retain such young women in school beyond the compulsory years.

Young women need careful mentoring and career advice at school especially in Years 11, 12, and 13 so that they maximize their options for the future. They need encouragement to become less dependent learners and they need to become less liable to take softer options when these are not in their best interests. Indications are that more course-work assessment encourages female achievement, and probably New Zealand should be altering its assessment patterns. New Zealand society needs to award more value to those aspects of mathematics that young women excel in, and society, especially female society, must support young women who wish to study mathematics.

REFERENCES

Blithe, T., Clark, M., & Forbes, S. (1993). *The testing of girls in mathematics*. Wellington: Victoria University.

Burton, L. (1993). *Differential performance in assessment in mathematics at the end of compulsory schooling*. Paper presented to the Fifth Conference of the European Association for Research on Learning and Instruction, Aix-en-Provence, September 1993.

Clark, M. (1993a). Friend or foe? The pressures for change. *New Zealand Journal of Mathematics, 22,* 29–42.

Clark, M. (1993b). *The effect of context on the teaching of statistics at first year university level*. Paper presented to the First International Conference on Statistics Education, Perugia, Italy.

Department of Education. (1987). *Mathematics achievement in New Zealand secondary schools. A Report on the conduct in New Zealand of the Second International Mathematics Study within the International Association for the Evaluation of Educational Achievement*. Wellington: Department of Education.

Forbes, S., Blithe, T., Clark, M., & Robinson, E. (1990). *Mathematics for all? Summary of a study of participation, performance, gender and ethnic differences in mathematics*. Wellington: Ministry of Education.

Forbes, S., & Mako, C. (1993). *Assessment in education: A diagnostic tool or a barrier to progress*. Paper presented at the 1993 World Indigenous Peoples Conference:

Education, Wellongong, Australia.

Keitel, C. (1993). Unpublished comment at International Commission on Mathematics Instruction Study Group on Gender and Mathematics, Höör, Sweden.

Knight, G., Arnold, A., Carter, M., Kelly, P., & Thornley, G. (1993). *The mathematical needs of New Zealand school leavers*. Department of Mathematics. Palmerston North: Massey University.

Ministry of Education. (1992). *Mathematics in the New Zealand curriculum*. Wellington: Ministry of Education.

Ministry of Education. (1993). *Education statistics of New Zealand 1993*. Wellington: Ministry of Education.

Morton, M. (1993). Doctoral studies in mathematics and statistics in New Zealand. *The New Zealand Mathematics Magazine, 30*(2), 42–45.

Pole, N. (1993). The challenge of demographic and labour market trends for the senior secondary school. *The research bulletin* (2). Wellington: Research and Data Management and Analysis Sections, Ministry of Education.

Purser, P., & Wily, H. (1992). *What turns them on: An investigation of motivation in the responses of students to mathematics questions*. Christchurch: Burnside High School.

Rivers, M., Lynch, J., & Irving, J. (1989). *Project FAST: Report of a national pilot study*. Wellington: Department of Education.

Sturrock, F. (1993). *The status of girls and women in New Zealand education and training*. Wellington: Ministry of Education.

Wilson, M. (1986). *A report on the status of academic women in New Zealand*. Wellington: Association of University Teachers of New Zealand.

Young-Loveridge, J. (1991). *The development of children's number concepts from ages five to nine*. Hamilton: University of Waikato.

RUI FEN TANG, QI MING ZHENG, AND SHEN QUAN WU

GENDER AND MATHEMATICS EDUCATION:
A SNAPSHOT OF CHINA

China was dominated by feudalism for several thousand years. The forces of tradition pressed heavily upon Chinese women under this regime. It can be said with certainty that there was no equality between the sexes. An old Chinese proverb said "for women without talent means virtues." Not surprisingly, women were deprived of an education.

Things have changed a great deal since 1949. "Equality between the sexes" and "Everybody has the right to education" were slogans energetically advocated by the government. A series of decrees were promulgated to improve the situation of women. Tables 1 and 2 provide a picture of women in education in present-day China.

Table 1: Student enrolment by level of education, 1993

	Total students	Female students	% Female
Kindergarten	2.43×10^7	1.14×10^7	46.9
Primary school	1.22×10^8	5.69×10^7	46.6
Junior high school	4.07×10^7	1.78×10^7	43.7
Senior high school	7.05×10^6	2.75×10^6	39.0
Junior vocational	5.64×10^5	2.30×10^5	40.8
Senior vocational	2.86×10^6	1.35×10^6	47.2
University & college	2.18×10^6	7.36×10^5	33.8
Graduate	9.42×10^4	2.34×10^4	24.8

From Table 1 and Table 2, we can see that the percentage of females both as students and as teachers decreases as the level of education increases, reflecting how still very entrenched the traditional ideas about women and education are in the minds of the Chinese. Girls' rights to an education are still denied in some regions, particularly in the countryside.

As in most countries, the difference in mathematics achievement between girls and boys is significant in China. In general, the performance of girls in mathematics is not satisfactory. Is this an unchangeable law? Are there insuperable barriers to girls' mathematical achievement? Can we do something to change the situation and improve the performance of girls?

G. Hanna (ed.), Towards Gender Equity in Mathematics Education, 271–276.

Table 2: Teachers by level of education, 1993

	Total teachers	Female teachers	% Female
Kindergarten	8.15×10^5	7.70×10^5	94.5
Primary school	5.53×10^6	2.46×10^6	44.5
Junior high school	2.56×10^6	8.88×10^5	34.7
Senior high school	5.76×10^5	1.58×10^5	27.4
Junior vocational	3.20×10^4	7.58×10^3	23.7
Senior vocational	2.16×10^5	7.59×10^4	35.1
University & college	3.88×10^5	1.16×10^5	29.9

SHANGHAI WEI YU EXPERIMENT

Shanghai Wei Yu Secondary School has a long history. Six grades are taught at the school. Since girls make up 40 percent of the student population, any efforts made to improve the mathematical achievement of girls affect a large sector of the student group. In recent years, mathematics teachers in this school have undertaken several measures in this regard.

First of all, the teachers have adopted a non-sexist attitude in the teaching of mathematics to girls. In the past, some teachers often assumed that while girls can learn foreign languages successfully, they cannot learn mathematics well, especially the mathematics of the more senior grades. Despite these perceptions, girls do get good results on various examinations and in mathematics competitions. More girls than boys achieved a grade of 93% or greater on the college entrance examinations in 1993. Tables 3 and 4 present the marks obtained in the regional mathematics competition in 1990 (the highest possible mark was 120).

Table 3: Achievement of one class in regional
mathematics competition, 1990

	Number entered	Mean mark	Highest (lowest) mark	1st Prize winners	2nd Prize winners	3rd Prize winners
Girls	5	93.6	100(88)	1	1	2
Boys	5	88.8	104(68)	1	0	2

A second approach tried at the school is one of cultivating the girls' confidence in their ability to learn mathematics. It was thought that girls' representation in mathematics was tremendously influenced by psychological barriers (e.g., their sense of inferiority). For this reason, the following means were adopted:

Table 4: Achievement of whole grade

	Number entered	Mean mark	Highest (lowest) mark	1st Prize winners	2nd Prize winners	3rd Prize winners	% winners
Girls	15	85.87	100(68)	3	1	2	40
Boys	26	81.04	104(44)	3	1	2	24

- publicizing those girls who had improved their achievement in mathematics, highlighting their learning methods and positive experiences; this kindled the enthusiasm of other female students

- praising girls on their progress; failures often result in girls losing their confidence and may even cause them to be afraid of mathematics

- not excluding girls from competition; the results show that girls do get good results in many of the competitions, sometimes even surpassing the boys

- emphasizing the virtues of girls which are helpful in learning mathematics: careful and methodical work habits and a sense of seriousness about learning, while at the same time remedying their defects: a lack of flexibility and a relatively poor ability to synthesize

The third measure is to enhance individual coaching for girls and vigorously stress the foundations of learning mathematics. This has been attempted by:

- freeing girls of any misgivings they might have about mathematics, removing any psychological or ideological obstacles as much as possible, and instilling an enthusiasm towards mathematics

- coaching girls in accordance with their special, individual characteristics; every girl has different needs and requires an analysis of what is interfering with her mathematical learning and a determination of the means of reversing those interferences

- instructing girls in the methods of learning mathematics: encouraging them to develop an understanding of the subject instead of mechanically memorizing, to sharpen their ability to distinguish between concepts and ideas, and to enhance their flexibility and ability to synthesize

Through this series of measures, the girls' confidence in learning mathematics is heightened, their learning methods are improved, and their achievement in the subject is more probable.

The following tables demonstrate the achievements made in the college entrance examinations in 1989 and 1993 and show a comparison between the performance of girls and boys in one class (the highest possible mark was 150).

Table 5: Results of 1989 college entrance examinations,
Shanghai Wei Yu Secondary School

	Number entered	Mean mark	# (%) Obtaining > 115	# (%) Obtaining 110-114	# (%) Obtaining 90-99	# (%) Obtaining < 89
Girls	20	104.15	7 (35%)	6 (30%)	4 (20%)	3 (15%)
Boys	23	105.01	8 (35%)	9 (39%)	2 (9%)	4 (17%)

Table 6: Results of 1993 college entrance examinations

	Number entered	Mean mark	# (%) Obtaining > 120	# (%) Obtaining 110-119	# (%) Obtaining 90-109	# (%) Obtaining < 89
Girls	14	11·. 6	7 (50%)	5 (36%)	2 (14%)	0 (0%)
Boys	23	115.26	9 (39%)	11 (48%)	2 (9%)	1 (4%)

Reflection

Since the experiment at Shanghai Wei Yu Secondar, …. ɔl was carried out by only one teacher in one school and the chosen san. was relatively small, the results may be considered specific and incidental. However, the outcome of any experiment has significance, and a good deal of enlightenment can be gained from this one.

First, these results give a positive answer to the question "Are girls capable of learning mathematics successfully?" Girls can achieve not only in the lower grades but also in the higher ones and they can perform well in mathematics competitions.

The results of the experiment verify again the simple premise that there are no physical or intellectual barriers to the participation of girls in mathematics, science, or technology; girls' inferiority in mathematics cannot be explained on the basis of biological differences. Therefore, we must find the source of the gender differences in mathematics achievement.

Second, there are various factors which make girls avoid mathematics and related fields: social and cultural barriers (i.e., the bias that exists in the larger society and in particular the attitudes of teachers and parents), the influence of tradition, and the consequences of these, especially the psychological impact on girls' beliefs and confidence.

In contrast with women in other countries in the world, it seems that Chinese women are burdened with a heavier pressure — the remnants of feudal thinking — which has not yet been entirely overcome. Many people, including teachers, parents, and girls themselves, still believe that men are superior to women in many respects. Most teachers in China attribute boys' poor achievement to inadequate effort but girls' to poor ability, and girls are apt to be influenced by their teachers' assessment more than boys.

Finally, the Shanghai Wei Yu experiment also shows that success is the fruit of hard work and does not fall from the skies. In order to change gender discrimination, there is a lot that needs to be done. To carry out this very important and difficult task, we need to focus on several tasks:

1) conducting a propaganda campaign among the masses to mold public opinion and dispel the negative influence of traditional ideas and historical prejudices about gender and mathematics education. Equality between the sexes is a perfectly justifiable goal and everyone needs to correct their thinking in order to realize it.

2) changing teachers' ideas, since teachers are one of the most important educational factors affecting students; to a certain extent teachers have a greater influence on students than parents. Teachers' treating girls and boys as equals would probably result in improved mathematical performance on the part of girls.

3) consistently encouraging girls to foster a realistic self-evaluation and a positive attitude and to take an active approach in dealing with the matter of confidence. Internal factors, psychological makeup, play a dynamic role in influencing students' achievement in mathematics.

4) adopting and enhancing specific measures — for example, analysing girls' work habits and identifying their strong and weak points. In the meantime, we should encourage girls to build on their achievements and correct their mistakes to do even better in the future. Of course, teachers need to fit the remedy to the cause, giving individually designed support and coaching to girls.

SUMMARY

An upsurge in economic development is bound to be followed by an upsurge in the cultural field. There has been, in this century, a rapid development in mathematics education alongside the tremendous growth in science and technology. As a result, more and more people are encountering far more mathematics than ever before. In China, as in most countries of the world, students are expected to learn a considerable amount of mathematics. Thus, the importance of girls' achievement in mathematics should be recognized.

Although it has been demonstrated that girls are just as capable of successfully learning mathematics as boys, administrative authorities must help break down traditional stereotypes by implementing effective measures to ensure the due right of girls in all respects. Teachers must also devote attention to this by having a positive attitude, finding feasible means to enhance girls' interest and self-confidence, and by giving special help to girls so they can improve their learning. These requirements are appropriate for parents as well.

Girls themselves should strive to do their best, regard themselves highly, make unremitting efforts to improve themselves, and conduct themselves with dignity. They have to take the initiative to acquire legal status and all rights in every respect. Girls should not underestimate their abilities, especially their level of mathematics performance. At the moment, it may be necessary to have affirmative action plans for girls, since the past has favored boys.

Through extensive study and research and intense communication and discussion, it is hoped that the unjustifiable phenomenon of gender discrimination will be eliminated in the near future. However, it will likely need the efforts of several generations to realize this worthy objective.

CARLOS BOSCH AND MARIA TRIGUEROS

GENDER AND MATHEMATICS IN MEXICO

Mexico is a poor, underdeveloped country. The social structure is characterized by well-defined classes that differ not only in economic but also in cultural terms. Although lately Mexico has been in the news thanks to new
economic policies, the social structure has not changed much, and economic differences are wider than ten years ago.

As we are concerned here with education, we will deal with class differences from a cultural point of view. People in the lower class are mostly
illiterate. They have only three or four years of schooling and scant
resources. Gender differences are strong; men have much more freedom
than do women. Women are in charge of the household but also have to
work. They are oppressed, and many of them are abandoned by their husbands, receiving little or no help. In the middle class, one finds people who
have studied at the elementary and high school levels, even some who have
a bachelor's degree. In economic terms they are better off, but there is, in
general, a lack of interest in cultural activities and in reading. Women in
this class, depending on their economic status, work outside the home or
dedicate themselves to their home and children. Gender differences are not
so great as in the lower classes, but still men enjoy much more liberty than
do women, as women are supposed to take charge of everything at home
without any help whatsoever. The higher cultural class is not necessarily
equivalent to the higher economic class. In the former, we find people who
are well educated and who are concerned about all kinds of cultural activity. Although women in this class are treated more as equals, have good
jobs, and enjoy more freedom, there still exists a certain degree of oppression. They are still in charge of everything at home, although occasionally
they receive some help from their husbands.

Family values and social expectations determine gender roles in very
definite ways, to the point where a general "cult" of the mother figure
exists. Socially, the women's role is very important because it implies the
stability of the family. This has an important influence on career choice.
Well-off families are willing to accept one of their daughters studying
towards a math-related career because they know her future household
won't depend economically on her. They are more opposed to their sons
doing the same.

G. Hanna (ed.), Towards Gender Equity in Mathematics Education, 277–284.
© 1996 *Kluwer Academic Publishers. Printed in the Netherlands.*

EDUCATION IN MEXICO

In the past ten years, the Mexican educational scene has undergone a dramatic change. The university population has grown enormously. (See Figure 1.) Acquisition of a degree means better economic and social opportunities. Everybody wants to have a degree. Public university is free, and once students are enrolled they can stay almost as long as they wish. Government policies favor mass education and democratic access to university programs.

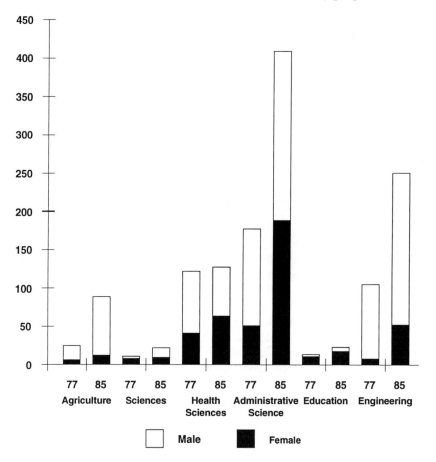

Figure 1: Distribution of college students' enrolment by
sex and subject area (1977–1985)

This growth of the student population has been accompanied by very rapid growth in the number of full-time professors and by the diversification of specialties. (See Table 1.) In a recent study (Gil-Anton, no date), researchers found that in the past ten years, an average of 12 professors per day had to be recruited to satisfy student demand. (See Table 2.) This bears comparison with growth in other countries.

Table 1: Distribution of faculty by subject area

Area	up to 1959	1960-1969	1970-1985	1986-1992	1992 only
Agriculture	2.8%	3.7%	4.0%	4.3%	4.0%
Health sciences	32.9%	27.9%	23.2%	14.5%	21.8%
Sciences	9.9%	16.7%	11.9%	9.6%	11.6%
Social and administrative sciences	15.7%	17.7%	28.9%	37.0%	29.7%
Education and humanities	11.9%	5.8%	7.1%	8.3%	7.2%
Engineering and technology	26.9%	28.2%	25.0%	26.3%	25.6%
Total number of faculty (100%)	55	350	2,251	867	3,522

Table 2: Expansion of academic positions

Years	Absolute increment	Relative increment	Average positions per day
1960–1970	14,307	133.1%	4.36
1970–1985	70,733	282.3%	12.92
1985–1989	8,418	8.8%	5.77
1989–1992	9,031	8.6%	8.25
1960–1992	102,489	953.5%	8.77

In the past ten years, the average recruitment of full-time university professors was one per day in Belgium and eight per day in West Germany. There has been a marked increase in university faculty in Mexico, but the academic structure is not as well organized there as in Germany; the arrival of new faculty caused a great deal of confusion at every level: academic, administrative, and social, resulting in low salaries and a concentration of ill-prepared teachers. Explosive growth caused a rift in the educational system. Most of the work that is being done currently is directed towards reorganization of the system. Critical studies of the situation are scarce; however, and this lack of information makes reorientation difficult.

A careful analysis of university faculty reveals some striking features. Two-thirds of the faculty come from families where neither of the parents has higher education. (See Table 3.) More men than women are in this situation. This implies significant social mobility, but in terms of general culture and preparation, these teachers tend to be ill-equipped and narrow-minded. Almost one-third of the faculty were recruited before receiving their bachelor's degree and without any professional and teaching experience. (See Table 4.) In the sciences and mathematics, two-thirds were in this situation although half of them had some teaching experience. (See Table 5.)

Table 3: Distribution of faculty by subject area and
level of parents' education

Area	None of the Parents Went to College	At Least One of the Parents Went to College
Agriculture	67.4%	32.6%
Health sciences	69.6%	30.4%
Sciences	71.7%	28.3%
Social and administrative sciences	69.7%	30.3%
Education and humanities	67.5%	32.5%
Engineering and technology	67.0%	33.0%

Table 4: Distribution of faculty by subject area and educational level
at the beginning of their first position

Area	No BA	BA	MA	PhD	Others
Agriculture	29.8%	54.9%	10.7%	4.2%	0.4%
Health sciences	22.2%	56.3%	3.7%	0.9%	16.9%
Sciences	60.3%	33.9%	4.1%	1.7%	0.0%
Social and administrative sciences	35.4%	56.5%	4.3%	2.0%	1.3%
Education and humanities	42.7%	42.6%	9.2%	3.8%	1.8%
Engineering and technology	38.0%	54.3%	4.7%	1.4%	1.6%
Total	36.4%	52.1%	4.9%	1.8%	4.7%

In contrast, currently only about 4 percent of the faculty are recruited under these conditions. In the sciences and mathematics, most of the faculty have at least a master's degree and do some research, in contrast to the social and administrative sciences where most of the faculty members have only a bachelor's degree and only 16 percent do some research. When asked what they would like to do to improve their qualifications, scientists and mathematicians demanded courses in their areas to improve their knowledge, while

Table 5: Educational level of faculty reached in 1992 by subject area

Area	No BA	BA	MA	PhD	Others
Agriculture	2.8%	48.8%	30.2%	5.1%	13.1%
Health sciences	1.0%	44.5%	14.2%	3.2%	37.0%
Sciences	3.9%	43.7%	32.4%	18.0%	2.0%
Social and administrative sciences	4.8%	60.8%	17.7%	3.2%	13.5%
Education and humanities	11.1%	49.3%	22.0%	10.5%	7.1%
Engineering and technology	5.1%	65.9%	16.5%	2.0%	10.4%
Total	4.3%	55.3%	19.2%	5.2%	16.0%

administrative and engineering faculty members asked for teacher training.

There are more female professors in private than in public universities. This means that there are more women who teach than do research, because private universities have few research projects. A recent survey reveals that more than half of the women faculty members prefer to work at the university; only about 40 percent would like to be working elsewhere. This may be related to the feasibility of working while taking care of their children and home. More than half of these women live solely on their university salaries, in contrast to just 33 percent of the men. The male professors generally have other projects that complement their salaries.

In contrast to this expansion in the university, the elementary level of education has grown more slowly. Devaluation of the teacher's job and the low purchasing power of their salaries has resulted in a decline in the numbers of well-prepared teachers and of students in the area of education. One fact worth noting is that although government policy statements are always addressed to male teachers, 80 percent of elementary school teachers are women. It would be very interesting to investigate whether this has any consequence in the way children approach mathematics and if this could account for the low achievement of Mexican children in mathematics.

When one looks at students, those who take degrees related to mathematics are headed for the engineering field. There is no tradition in the sciences and mathematics. These careers are almost unknown. Students think that mathematics is only a tool for problem solving, useful only for applications. They don't think of it as a discipline and aren't aware of what a mathematician or a scientist does. Maybe this is because these areas were instituted in Mexican universities only about 50 years ago, but we think instead it is due to the way in which students are taught in elementary and high school, and because popularization of science is very limited.

In 1969, a study by Liliana Morales (1969) found that the percentage of women students in higher education was 17.3 percent; in 1985 women made up 34.5 percent. The student population grew 4.19 times in that period. The number of male students grew 3.339 times and that of women 9.4. These figures show the importance of the participation of women in the accelerated expansion of higher education in the last 15 years.

This expansion gives reason for optimism, as it can be considered to demonstrate progress towards equal opportunity for the sexes. It can also be viewed less optimistically. The Mexican educational system is highly selective and further analysis of data shows that, although the number of girls in elementary education is the same as the number of boys, girls have a higher school dropout rate than do boys. Gender inequity in access to education is closely related to the economic development of geographic areas in the country, being higher in urban areas and lower in rural areas where the elementary school system is deficient. When one looks at the distribution in the different subject areas, one finds that in the sciences and mathematics the number of students has doubled, but the percentage distribution of 63.2 percent men and 36.8 percent women did not change from 1977 to 1985. Engineering and technology show an increase in female participation from 8 to 14 percent. Whereas in the health sciences, the female population increased by 28 percent — the same percentage as the decrease in the male population (in this area there has not been any overall growth since 1977) — in administrative science we find an increase from 32.7 percent to 45.2 percent in female participation and the striking result that, of the total increase in university women, 65 percent were absorbed by this area. In education and the humanities, there are still more women students, but it is interesting to note that the percentage in 1977 was 59 percent, whereas now it is 57 percent (Anuarios Anuies).

In general terms, we can see a real advance in the democratization of education, but it is still very unequal in different areas, with math-related disciplines behind the others. It is worth noting that women are entering in higher education in greater numbers but it seems they are pursuing highly competitive careers with fewer possibilities in the marketplace, as those careers are being abandoned by men.

GENDER AND MATHEMATICS

As one can see, there are important problems in the Mexican educational system and specifically in math education. In a country of 80 million people we have about 1,800 bachelor's, 600 master's, and 250 doctoral degrees in mathematics, including those with a teaching emphasis. These few mathematicians take care of education, research, teacher preparation, and everything else related to mathematics.

Government policies in education are characterized by equal gender opportunity, so we think social issues have a direct impact on problems related to gender and mathematics education.

The lack of information about gender and education makes it difficult to address the problem. We, therefore, based the following considerations on the analysis of interviews of women involved in the field of mathematics, students in their last two years of high school, and the parents of these students.

Parents play an important role in career selection. Not only is there a relationship between parents' education (specifically that of the mother) and the access of women to higher education but social class also determines the amount of encouragement parents provide for their children to study mathematics. Only in the cultural upper class can one find parents who don't hold stereotypical views of women in mathematics. The general pattern is one of a lack of parental encouragement to enrol in advanced mathematics courses.

There is a widespread opinion that mathematics-related careers are careers for men. There is also a lack of economic incentive for pursuing these careers. So parents often would tolerate more easily a daughter studying sciences or mathematics than a son, due to the low salaries.

The Mexican Mathematical Olympiad is a program designed to popularize mathematics. It started in 1987, and in six years there have been only two girls among the 36 winners. Even with these results, the percentage of women in the national competition has increased from 9 percent in 1987 to 28 percent in 1992. Parents play in important role in this difference in participation, because they don't like their daughters to travel by themselves or with male students and they hesitate when girls ask permission to participate in a math competition.

Mathematicians are aware of the gender problem and are taking action to ensure that there is gender equity in new elementary school textbooks — in the illustrations and in the different activities proposed. The Ministry of Education, which is responsible for all curriculum change, recently held an open contest for mathematics textbooks. A special commission selected the best entries. One of the criteria for selection suggested implicitly by the Ministry was gender equity.

There is a widespread belief across all social sectors that achievement in mathematics is a clear sign of intelligence. While this belief is a stimulus for male students to continue studies in mathematics and sciences as a key factor for success in the job market and their social role, it deflects women away from mathematics and the sciences. Intelligent women have problems in their relationships with men and are isolated from their peer groups. In making career choices, girls are vulnerable to peer values. They select their careers together with their girlfriends. Although there is not a clear difference in performance in mathematics up to high school, social beliefs and the role played by women in society, as outlined above, inhibit the choice of math-related careers for women.

Women who choose these careers are usually good achievers and are good students but their research performance is lowered by the multiple role they have to play in family and society. There are more women than men who drop out from the National Research System because they are unable to maintain the required publication rhythm.

SUCCESS AND FAILURES

There is a lack of research on the problem of gender in mathematics in Mexico. Although there is a society of women in sciences, its impact has not reached beyond the group of women scientists themselves. There is not widespread awareness of the problem and this makes its solution difficult.

The Ministry of Education is concentrating its efforts and policies on the restructuring of education in general. Mathematics education is a topic of concern but gender inequities are not on the agenda. Teacher-training programs don't address gender inequities, and as teachers mirror social beliefs and prejudices, there is a tendency to reinforce them rather than to address and fight against the prevailing social beliefs. Awareness in the mathematics community and on the part of members of the National Program for Teacher Training brings hope for change in this situation, as exemplified by the new textbooks.

The popularization of science is growing in Mexico and can play an important role in changing the attitudes towards mathematics. Some applied mathematics programs oriented to economics and social sciences are recruiting more women and directing them successfully to graduate schools in foreign countries.

We think that changes in the educational system in general — its growth, and its tendencies towards democratization of both student populations and university faculty — will continue. This, together with the close relationship between parents' education, particularly the mother's, and children's education, will reinforce the next generation. We expect future generations to be balanced in terms of gender and to be better prepared. So when one thinks of the long-term effect of the system's expansion it can be viewed as a success.

REFERENCES

Morales, L. (1969). *La mujer en la educacion superior de Mexico*. Universidad Futura num.1.UAM-A 1969. Mexico D.F.

Anuarios ANUIES 1977-1985.

Gil-Anton, M. et al. (no date). *Los rasgos de la diversidad*. UAM-A 1994, Mexico D.F.

SUSAN F. CHIPMAN

FEMALE PARTICIPATION IN
THE STUDY OF MATHEMATICS:
THE US SITUATION

In the United States, mathematics is the least sex-typed of major fields of study at the university level. That is, women have been about as well represented among the recipients of bachelor's degrees in mathematics as they have been among the recipients of all bachelor's degrees. This has been true for many years. In 1950, the first year for which complete statistical records on degrees awarded in the US (National Center for Education Statistics, Series of Earned Degrees Conferred) are available, 24 percent of all BA degrees were awarded to women and 22.6 percent of the BAs in mathematics. By 1977, there had been a great increase in women's participation in US higher education: 46.1 percent of all BAs were awarded to women in that year and 41.5 percent of the BAs in mathematics. By 1985, women had attained a slight majority (50.7%) among all recipients of BAs and received 46.1 percent of the BAs in mathematics. In the most recent year for which data are available, 1991, women received 54.1 percent of all BA degrees awarded and 47.2 percent of the BAs in mathematics. There is no other field of study which comes so close to proportional representation of men and women. Psychology is perhaps the closest contender, but in recent years it has become predominantly female, like humanities fields; social science fields are predominantly male. Engineering and physical science fields, of course, remain predominantly male.

These facts about the nearly proportional representation of women in university-level mathematics come as a great surprise to most people in the US. No doubt they are also very surprising to those in other countries who are aware of the great attention that has been given to the "issue" of women's participation in mathematics by educational researchers in the US. Certainly the high rate of participation in the study of mathematics itself casts doubt on the widespread hypothesis that an aversion to mathematics is responsible for the low rates of participation of women in many fields that are seen as having a substantial mathematical component — engineering and physical sciences. If mathematics were the source of the problem, one would expect to see the lowest rate of participation in mathematics itself.

Of course, mathematics itself is a rather small field of study, in 1991 accounting for only 1.3 percent of all US BAs, down from 3 to 4 percent in earlier decades. Excellent participation in mathematics at the university level would not necessarily be incompatible with substantial differences in secondary school participation.

G. Hanna (ed.), Towards Gender Equity in Mathematics Education, 285–296.
© 1996 *US Government. Printed in the Netherlands.*

SECONDARY SCHOOL PARTICIPATION IN
MATHEMATICS STUDY

At one time, there were very substantial gender differences in secondary
school mathematics course enrolments in the US, where the study of mathematics is usually optional after the first two years. A nationally representative survey sample that was taken in 1960 showed that 9 percent of the girls,
as compared to 33 percent of the boys, were taking four years of high school
mathematics (Wise, 1985). By the time the comparable National Longitudinal Sample was taken in 1972, these figures had changed to 22 percent and
39 percent (Wise, personal communication). Data from the College
Entrance Examination Board (CEEB) suggest that gender differences in
secondary school mathematics study have become very small in recent
years, at least for that large segment of the population which has an interest
in attending colleges or universities. CEEB's "College-Bound Seniors: 1993
Profile of SAT and Achievement Test Takers," for example, indicates that
51.8 percent of those who reported studying four or more years of secondary school mathematics were female while 54.2 percent of those providing questionnaire responses were female. Forty-seven percent of those who
reported taking calculus in secondary school were female. Obviously, the
large gender differences in mathematics enrolment have disappeared. This
great change over the last 30 years can probably be attributed to the coinci-
dent change in expectations that women would pursue lifelong employment
and their correspondingly increased participation in university-level education. US colleges and universities, especially those which are somewhat
selective and desirable, typically expect or require four years of high school
mathematics for entrance. In the last 20 years, the most selective institutions
have come to expect the study of calculus to be included in those four years,
and the better secondary schools have arranged their curricula so that calculus will be included in the standard four years of mathematics taken by college-bound students. That is, female secondary students in the US are now
studying advanced secondary school mathematics to about the same extent
that male students do because that course of study is effectively required in
order to fulfil their intention to go on to college or university study. Back in
1960, the intention of going to college — as opposed to the intention of pursuing a, possibly, brief career in office work — was strongly predictive of
enrolment in advanced secondary school mathematics courses (Wise, 1985),
presumably because of college and university entrance requirements. Ironically, much of the change in female participation in mathematics study
seems to have occurred before the issue of female participation generated so
much concern; that concern itself seems to have been just one aspect of a
very broad social change that was going on during those years.

   Typically, secondary school students in the US do not have any specialization for particular fields. Such personal preferences may be expressed
only in the selection of a very small number of elective courses. Typically

also, students are not admitted to any particular course of study at a college or university: entrance requirements or expectations do not vary much in relation to the student's expected field of specialization. Consequently, female rates of participation in advanced high school mathematics seem to have been driven upwards towards equity by the de facto requirements for university entrance, even though female rates of participation in many of the university major fields seen as requiring mathematics (engineering, physical sciences) remain low. We can no longer argue that the failure to study mathematics in high school is a barrier to women's entrance into these fields; it is more likely that women are simply not very interested in studying these fields, for whatever reason. Mathematics, in contrast, seems to be about as appealing to women as it is to men.

MATHEMATICS ACHIEVEMENT

Enrolment in mathematics courses is not, of course, the only variable of interest in mathematics education. Achievement may vary among students enrolled in the same course. In the US and elsewhere, there is widespread belief that males outperform females in mathematics. However, the data for the US do not necessarily support this belief (Chipman & Thomas, 1985; Hyde, Fennema, & Lamon, 1990). Large, representative studies of US student populations have tended to find little or no gender difference in overall mathematics performance prior to the secondary school years, when the study of mathematics often becomes optional in the US. It has sometimes been reported that girls perform better on computational items, and sometimes that boys perform better on some types of word problem items (e.g. Marshall, 1984), but a recent meta-analysis concluded that no such differences are evident prior to secondary school (Hyde, Fennema, & Lamon, 1990). An exception to the general picture of equality is that searches for mathematically talented youths have generated reports that extremely high levels of mathematical performance on the Scholastic Aptitude Test (a test normally used for college entrance) at a young age (about 7th or 8th grade) are much more frequently found in males (Benbow & Stanley, 1980, 1983). These reports have received much publicity and have had a substantial effect on public beliefs about male and female performance in mathematics (Eccles & Jacobs, 1986). Because of the way in which these searches are conducted — that is, using methods which do not ensure representative sampling — it is difficult to know what one should conclude about the actual incidence of high levels of mathematics performance among young male and female students in the US. The US National Assessment of Education Progress (NAEP) (National Science Board, 1993), which does aim at achieving a nationally representative sample, does report that .2 percent of females at age 13 and .5 percent of males at age 13 attained the highest category of mathematical proficiency (p. 232) — characterized as involving multi-step problem-solving and algebra, as well as various other mathematical content usually taught during high school. (Traditionally, algebra was taught in the 9th grade in the US, when students are usually about 14 years old, but quite a few students now begin the study of algebra earlier. There may be gender differences in that early course-taking.) However, it is clear that the generalization from these reports to beliefs about the performance of more typical male and females students is not justified. In the general population, gender differences in mathematics performance prior to secondary school are negligible.

By the end of secondary school, however, gender differences in mathematics test performance that favor males have usually been reported in the US, and the performance differences seem to arise from problem-solving tests or items (Hyde, Fennema, & Lamon, 1990). For many years, the possibility that differences in course taking might account for these differences in mathematics test performance seems to have been ignored. Obvious as that hypothesis might seem, many seem to have concluded that such test results implied lesser mathematical ability among female students. Analysis of data from the representative survey sample of US students that was collected in the early 1960s — when there were very substantial gender differences in secondary school mathematics course enrolments — showed that course enrolments statistically accounted for nearly all of the gender difference in mathematics performance at the end of secondary school (Wise, 1985). However, as noted above, gender differences in course enrolments have diminished greatly. Nevertheless, a performance difference of about .4 standard deviation on the Scholastic Aptitude Test, a widely used but non-compulsory test for college and university admission, remains. Similarly the NAEP (National Science Board, 1993) results show that 5.6 percent of 17-year-old females but 8.8 percent of 17-year-old males were attaining the highest category of proficiency in 1990 (p. 233). On the other hand, the mean results for 17-year-old males and females in 1990 showed no difference (pp. 7, 231), the culmination of a gap-closing trend.

The picture is further complicated by the fact that different measures of mathematical performance yield different messages about gender differences. In an important review paper, Kimball (1989) showed that course performance measures consistently favor females. Similarly, the results of some studies tend to indicate that examinations which are closely tied to the instructed curriculum, like the New York State Regents Exam (Felson & Trudeau, 1991) or International Association for the Evaluation of Educational Achievement (IEA) Math content which is well-represented in the "implemented curriculum" (Hanna, 1989), are more likely to favor females. This side of the story is further reinforced by another recent, large-scale study drawing from a huge sample of students gathered to investigate the validity of the SAT exam. When males and females were matched by university math course taken (in the same institution) and by performance grade received in the course, it was found that the females had received scores nearly 50 points lower on the SAT exam (Wainer & Steinberg, 1992). In other words, the SAT underpredicted the performance of females relative to males in these mathematics courses, advanced as well as introductory. Such results suggest that the observed gender differences in performance on tests like the SAT may reflect differences in responding to the testing situation itself (Becker, 1990) or that they may arise from extracurricular differences in experience that are related to the content of some such tests.

POSSIBLE SOURCES OF GENDER DIFFERENCES
IN TEST PERFORMANCE

There are a variety of factors which might contribute to persisting gender differences in math test performance. Significant gender differences do remain for course enrolments in "extra" mathematics courses like statistics and probability that are not part of the standard college preparatory curriculum. Large gender differences exist in enrolments in courses like high school physics that may provide considerable practice in solving mathematics problems. If the majority of secondary school females are taking high school mathematics simply to fulfil college or university entrance requirements rather than to prepare for further study and careers which intrinsically require mathematical competence, their degree of involvement with the subject matter may be less than is common for males taking the same courses.

Studies which have attempted to analyse gender differences in performance on the SAT or similar tests have not yielded any great insight (Chipman, 1988). Individual test items can be found that show very large gender differences, but the reasons for those differences are not obvious. There has been little or no consistency in the apparent nature of such items from one study to the next. Occasionally, these analyses have appeared to confirm hypotheses that gender differences might be concentrated in items with geometric or spatial content, but this has not proved consistently true. (Furthermore, the belief in gender differences in spatial ability is also rather weakly supported by the actual evidence — see Linn & Petersen, 1985). Many have hypothesized that the stereotypically masculine content of mathematics word problems might account for some of the sex differences in performance on such items. Subjective examination of items that do and do not show large gender differences does not provide obvious support for that view. Furthermore, a recent experimental study (Chipman, Marshall, & Scott, 1991) did not confirm the popular hypothesis that sex-stereotyped content of math word problems would affect performance. In several studies of the SAT, it was found that a class of items ("data sufficiency items") which ask whether sufficient data are available to answer the question did consistently favor females. It is rather hard to say why such items should have favored females; one can speculate that perhaps females tended to actually attempt problem solutions, thereby improving their ability to answer these items correctly while consuming extra time that may have hindered their performance on the rest of the examination. In any case, the historical fact that The Educational Testing Service chose to drop a class of items that consistently favored females from a test that consistently favors males should cast doubt on the tendency to treat the SAT as if it were some gold standard of mathematical ability. The SAT is a timed, multiple-choice test that rewards test-taking strategies such as guessing based on partial information. The characteristics that produce good performance on the SAT are undoubtedly somewhat different than the characteristics that result in good performance in mathematics courses or mathematical work itself. Does the SAT deserve more weight than the grading judgments of college mathematics professors? Probably not. Unfortunately the gender difference in math SAT scores does disadvantage women in the college admissions process, in the award of scholarships, and perhaps in faculty attitudes and advice about course selection.

WOMEN'S PARTICIPATION IN THE GRADUATE STUDY OF MATHEMATICS

In contrast to the picture at the bachelor's degree level, statistics about female participation in the graduate study of mathematics in the US suggest that there may be a problem. Not surprisingly, female participation in education for advanced degrees of any kind has lagged levels of participation at the bachelor's degree level. One must take this into consideration when interpreting the percentages of degrees in mathematics that are awarded to women. Table 1 shows the relevant percentages.

Table 1: Percent of degrees awarded to women

	BA-All	Math BA	MA-All	Math MA	PhD-All	Math PhD
1950	24	23				
1960	35	27				
1970	43	37				
1975	45	42	45	33	22	10
1980	49	42	49	36	30	13
1985	51	46	50	35	34	15
1990	53	46	53	40	36	18

(Source: The primary source of these data is the Series of Earned Degrees Conferred, National Center for Education Statistics. Data 1950–1970 as cited in Chipman & Thomas (1985). Data 1975–1990 as cited in National Science Board, *Science and Engineering Indicators —1993*, Appendix Tables 2-19,2-25,2-27, pp. 272–285.)

As of the mid-seventies (Chipman & Thomas, 1985), I observed that women's share of bachelor's degrees in mathematics was about .9 times as large as their share of all BA degrees, whereas their share of master's degrees in mathematics was about .7 times their share of all master's degrees, and their share of doctoral degrees in mathematics was about .5 times their share of all doctoral degrees. Those ratios seem to have remained quite stable to the present time, indicating that there is some specific problem with the progression of women into the graduate study of mathematics.

Some other sources of data suggested the same conclusion. Although women mathematics majors in the 1972 National Longitudinal Study had a good record of BA completion, only 6.8 percent of them were in graduate study of mathematics in 1976 and 87.3 percent were not students at that time. In contrast, 49.1 percent of male mathematics majors had continued into the graduate study of mathematics and only 23.6 percent were not in some type of graduate study. The record of female mathematics majors contrasted unfavorably with that of female physical science majors, who were almost as likely as their male counterparts to continue in graduate study of physical science or mathematics (38% vs. 39%). A study of persons taking the Graduate Record Examination at about the same time (Casserly & Flaugher, 1977) gave similar indications. Nearly twice as many white male as white female mathematics majors were registering for the GRE. In addition, proportionately fewer (46.5% vs. 58.8%) of the females were intending graduate study in mathematics. Twice as many females (19.7% vs. 9.1%) were intending to do graduate study in education. Again, the situation of female physical science majors appeared more favorable, a higher percentage (55.8%) intending graduate study in physical science and a much lower percentage (4.3%) going into education.

This situation appears to warrant further examination. Given that the locus of the remaining gender discrepancies is at the level of graduate education, the professional academic community in mathematics should be taking responsibility for further investigation and self-examination. It seems that many female mathematics majors are — or were — opting for immediate employment after completion of the BA degree. (For most, this employment probably was not teaching, because a master's degree of some type, typically in education, has been a usual requirement for teachers in the US for some years now.) It may be that female and male mathematics majors pursue somewhat different courses of study, corresponding to different career intentions. Although we know that female students do well in the mathematics courses they take (Kimball, 1989), we do not seem to know whether male and female recipients of the BA in mathematics are equally well-prepared for graduate study in mathematics. Data from the Graduate Record Examination (not necessarily a representative sample) show that women mathematics majors scored about 30 points lower on the quantitative section but equally high as men on the analytic section in both 1979 and 1984 (National Science Foundation, 1986). Mathematics departments should examine whether their male and female students are following different courses of study and, especially, whether they are being advised to follow different courses of study for reasons that may be dubious. In particular, the results of Wainer and Steinberg (1992) suggest that the SAT scores of female students should be interpreted more positively. Alternatively, it may be that women have been discouraged by the prospects for their employment after graduate study in mathematics. Historically, these prospects have not been good: one-third of the recognized creative women mathematicians studied by Helson (1971) were unemployed. In the mid-seventies, women with doctorates in mathematics were three times as likely as men to be unemployed. In more recent years, the employment prospects for mathematicians generally have not been good, and the total numbers of doctoral degrees in mathematics being awarded dropped substantially during the 1980s. US women may have estimated that they were unlikely to be rewarded for continuing into the advanced study of mathematics.

THE AFFECTIVE DIMENSION

Speculations about the effect that employment expectations might have on the decision to do graduate study in mathematics should remind us that affective variables, not just cognitive ones, influence participation in mathematics study. The largely mistaken view that US girls were not studying mathematics in secondary school inspired many studies of feelings about mathematics.

Two findings of these studies have been consistent and notable (Chipman & Wilson, 1985; Hyde, Fennema, Ryan, Frost, & Hopp, 1990). In the US at least, female students *like* mathematics as well as male students do. However, consistently, a small gender difference in confidence in oneself as a learner of mathematics is found. This gender difference is hardly surprising. There is a pervasive social stereotype that females are less capable in mathematics, achieve poorly in mathematics, and need special help in mathematics. Despite their dubious validity — as outlined above — statements reflecting these assumptions appear regularly in the US media. Ironically, many of these statements accompany stories about well-meaning efforts to assist female students, such as provision of special single-sex mathematics classes. Because of a study that they had ongoing at the time, Eccles and Jacobs (1986) were able to document that the original Benbow and Stanley (1980) report and accompanying barrage of somewhat distorted publicity had a negative impact on the expectations that both girls and their parents had for their achievement in mathematics. (Headlines at the time read: "Do males have a math gene?" [*Newsweek*], "A new study says males may be naturally abler than females," [*Time*], "Are girls born with less ability?" [*Science*].) A strong case could be made that the major gender problem in US mathematics education today is the gender difference in confidence driven by pervasive social stereotypes that fail to recognize the actual mathematical accomplishments of female students. It is for this reason that this chapter began by presenting the positive side of the story.

IMPACT OF MATHEMATICS CONFIDENCE

Because strong correlations exist between measures of mathematics achievement (or other measures of intellectual ability) and measures of mathematics confidence or anxiety, it has been difficult to tease out the effects of the gender difference in confidence on students' educational and career choices. A number of results in the studies reviewed by Chipman and Wilson (1985) strongly suggested that mathematics confidence or anxiety was affecting student decisions about enrolling in advanced secondary school mathematics. A recent study of the influence of mathematics anxiety/confidence on the career interests and college major selections of Barnard College students (Chipman, Krantz, & Silver, 1992, 1993) was able to demonstrate a strong influence of mathematics anxiety/confidence, independent of the effects of quantitative SAT scores. In this select population (mean QSAT about 600), QSAT had no effect on expressed interest in a scientific career at the time of college entrance, whereas mathematics anxiety/confidence did have a significant effect. For actual biological science majors, the same picture held true at the end of college: no effect of QSAT, significant effect of mathematics anxiety/confidence. For actual physical science majors at the end of college, both QSAT and mathematics anxiety/confidence showed significant effects on the choice of major. In this population, there were some individuals with very high QSAT scores and low mathematics confidence. The results of this study indicate that the gender difference in mathematics confidence may well be partially responsible for some of the under-representation of women in science and engineering fields. In the Barnard study, the full impact of that effect was expressed *prior to* college entrance. In addition, however, one must remember that there are very substantial gender differences in general interest patterns (such as interest in "things" as contrasted to "people") that predict such vocational choices.

Given these results, it remains somewhat surprising that women are so well represented among the recipients of the BA in mathematics. Quite possibly, however, gender differences in confidence and/or the associated pervasive social stereotypes have something to do with the much decreased representation of women among the recipients of the doctoral degree in mathematics. It may be that women are still being discouraged — or at least not actively encouraged — from the pursuit of graduate study. Research focussed on younger students has not provided strong evidence that teacher encouragement or discouragement or style of classroom interaction actually affects the outcome for female students (Chipman & Wilson, 1985), even though interaction patterns which seem to disfavor female students are often reported. Still, such factors may become more important at advanced levels where mentoring and apprenticeship education prevail.

OTHER MOTIVATIONAL FACTORS

External encouragement, internal confidence, and the expectation of eventual rewards in employment are among the many motivational factors that may influence persistence in the advanced study of mathematics. Competing interests and demands on the individual may be another. Although the study participants have now lived out their lives, and many social changes have occurred, the Terman study of gifted children (Terman, 1954; Terman & Oden, 1959) may suggest some other factors. For men in the Terman study, the *breadth* of interests was a negative predictor of career success, and women in the Terman study differed from men both in the direction of their vocational interests and in having broader interests. The culture of the US places a high value on being a well-rounded individual, and this continues to be even more true for women than for men. One study of attrition among female mathematics majors and female graduate students in mathematics (Maines, 1980; Maines, Sugrue, & Hardesty, 1981) at two US institutions found that female students of mathematics spent much less time on mathematics than did male math students. One would expect this to result in less accomplishment. It seems that female students are less likely to develop the intense, almost obsessive involvement with mathematics that may well be critical to truly outstanding achievement. But perhaps they are less involved because the community of mathematicians does less to recruit and involve them. Regardless, it is well to remember that putative gender differences in underlying mathematical abilities are not the only possible explanation for gender differences in the extremes of mathematical accomplishment — whether it be emergence in a Stanley talent search, exceptional achievement in the NAEP exams, or attainment of the doctoral degree. Effort, involvement, engagement, and mentoring come into play.

CONCLUSION

Within a relatively short span of years, we have moved from the presumption that women would never do anything in which mathematical knowledge was relevant to a situation in which — at least nominally — all fields are considered open to women. As late as the 1940s, females were barred from the study of advanced mathematics and physics in some US secondary schools. In the 1960s, the US National Science Foundation sponsored special summer programs for secondary school students to which female students were not admitted. As late as the 1950s, a large proportion of the women who were *recognized as creative mathematicians* in the US were unemployed (Helson, 1971). Against this background, it is remarkable that women students of mathematics are now doing so well. There is little reason to believe that remaining gender differences reflect innate differences in cognitive ability. US society, and the US mathematical community in particular, should avoid discouraging girls and women from the study of mathematics by suggesting that they are incapable of it. The objective record suggests that they are very capable, without any need for special treatment or sex-segregated classrooms. Since the under-representation of US women in mathematics now begins at the point of transition from undergraduate to graduate study, academic departments of mathematics should be examining themselves to determine whether their attitudes and advice to women students are responsible for this fact. It is not accounted for by the lower rates of participation of women in graduate study generally. Today the primary women and mathematics problem in the US may be that people keep talking about the women and mathematics problem.

REFERENCES

Becker, B. J. (1990). Item characteristics and gender differences on the SAT-M for mathematically able youths. *American Educational Research Journal, 27*, 65–88.

Benbow, C. P., & Stanley, J. C. (1980). Sex differences in mathematical ability: Fact or artifact? *Science, 210*, 1262–1264.

Benbow, C. P., & Stanley, J. C. (1983). Sex differences in mathematical reasoning ability: More facts. *Science, 222*, 1029–1031.

Casserly, P., & Flaugher, R. (1977). *An evaluation of minority student data from the 1975–76 GRE background questionnaire.* Unpublished manuscript, Educational Testing Service, January 14, 1977.

Chipman, S. F. (1988). *Word problems: Where test bias creeps in.* Paper presented at the annual meeting of the American Educational Research Association, New Orleans, April, 1988. (ERIC Document Reproduction Service No. TM 012 411).

Chipman, S. F., Krantz, D. H., & Silver, R. (1992). Mathematics anxiety and science careers among able college women. *Psychological Science, 3*, 292–295.

Chipman, S. F., Krantz, D. H., & Silver, R. (1993). *Mathematics anxiety/confidence and other determinants of college major selection.* Paper presented at the International Committee on Mathematics Instruction Study Conference on Gender and Mathematics Education, Hoor, Sweden, 7–12 October, 1993.

Chipman, S. F., Marshall, S. P., & Scott, P. A. (1991). Content effects on word problem performance: A possible source of test bias? *American Educational Research Journal, 28*, 897–915.

Chipman, S. F., & Thomas, V. G. (1985). Outlining the problem. In S. F. Chipman, L. Brush, & D. Wilson (Eds.), *Women and mathematics: Balancing the equation* (pp. 1–24). Hillsdale, NJ: Lawrence Erlbaum Associates.

Chipman, S. F., & Wilson, D. (1985). Understanding mathematics course enrollment and mathematics achievement: A synthesis of the research. In S. F. Chipman, L. Brush, & D. Wilson (Eds.), *Women and mathematics: Balancing the equation* (pp. 275–328). Hillsdale, NJ: Lawrence Erlbaum Associates.

Eccles, J. S., & Jacobs, J. E. (1986). Social forces shape math attitudes and performance. *Signs: Journal of Women in Culture and Society, 11*, 367–380.

Felson, R. B., & Trudeau, L. (1991). Gender differences in mathematics performance. *Social Psychology Quarterly, 54*, 113–126.

Hanna, G. (1989). Mathematics achievement of girls and boys in grade eight: Results from twenty countries. *Educational Studies in Mathematics, 20,* 225–232.

Helson, R. (1971). Women mathematicians and the creative personality. *Journal of Consulting and Clinical Psychology, 36,* 210–220.

Hyde, J. S., Fennema, E., & Lamon, S. J. (1990). Gender differences in mathematics performance: A meta-analysis. *Psychological Bulletin, 107,* 139–155.

Hyde, J. S., Fennema, E., Ryan, M., Frost, L. A., & Hopp, C. (1990). Gender difference in mathematics attitudes and affect: A meta-analysis. *Psychology of Women Quarterly, 14,* 299–324.

Kimball, M. M. (1989). A new perspective on women's math achievement. *Psychological Bulletin, 105,* 198–214.

Linn, M. C., & Petersen, A. C. (1985). Emergence and characterization of sex differences in spatial ability. *Child Development, 56,* 1479–1498.

Maines, D. R. (1980). *Role modeling processes and educational inequity for graduate and undergraduate students in mathematics.* Unpublished manuscript, Northwestern University, 1980.

Maines, D. R., Sugrue, N. M., & Hardesty, M. J. (1981). *Social processes of sex differentiation in mathematics.* Unpublished manuscript, Northwestern University, report on NIE-G-79-0114, 1981.

Marshall, S. P. (1984). Sex differences in children's mathematics achievement: Solving computations and story problems. *Journal of Educational Psychology, 76,* 194–204.

National Science Board. (1993). *Science and Engineering Indicators — 1993* ((NSB 93-1)). Washington, DC: U.S. Government Printing Office.

National Science Foundation. (1986). *Women and minorities in science and engineering* ((Report No. NSF 86-301)). Washington, DC: National Science Foundation.

Terman, L. M. (1954). Scientists and non-scientists in a group of 800 gifted men. *Psychological Monographs, 68*(7), 1–44.

Terman, L. M., & Oden, M. H. (1959). *Genetic Studies of Genius. The gifted group at mid-life* (Vol. 5). Stanford, CA: Stanford University Press.

Wainer, H., & Steinberg, L. S. (1992). Sex differences in performance on the mathematics section of the Scholastic Aptitude Test: A bidirectional validity study. *Harvard Educational Review, 62,* 323–336.

Wise, L. L. (1985). Project TALENT: Mathematics course participation in the 1960's and its career consequences. In S. F. Chipman, L. R. Brush, & D. M. Wilson (Eds.), *Women and mathematics: Balancing the equation* (pp. 25–58). Hillsdale, NJ: Lawrence Erlbaum Associates.

INDEX

AUTHORS' ADDRESSES

URSULA M. FRANKLIN
Department of Metallurgy and Materials Science
University of Toronto
Toronto, Ontario M5S 1A1
Canada

GILA HANNA
Department of Curriculum
Ontario Institute for Studies in Education
252 Bloor Street West
Toronto, Ontario M5S 1V6
Canada

ELIZABETH FENNEMA
Department of Curriculum and Instruction
School of Education
University of Wisconsin-Madison
Madison, WI 53706
USA

MARY GRAY
Department of Mathematics and Statistics
American University
Washington, DC 20016-8050
USA

GILAH C. LEDER
Graduate School of Education
La Trobe University
Bundoora, Victoria, 3083
Australia

HELGA JUNGWIRTH
Institut für Mathematik
Johannes Kepler Universität Linz
A-4040 Linz
Austria

SHARLEEN D. FORBES
Statistics New Zealand
P.O. Box 2922
Wellington
New Zealand

ANN HIBNER KOBLITZ
History Department
Hartwick College
Oneonta, NY 13820
USA

BARBRO GREVHOLM
School of Education
Lund University
Box 23501
S20045 Malmö
Sweden

KARI HAG
Department of Mathematical Sciences
The University of Trondheim, NTH
N-7034 Trondheim
Norway
and
Department of Mathematics
The University of Texas at Austin
Austin, TX 78712-1082
USA

BODIL BRANNER
Mathematical Institute
Building 303
Technical University
DK-2800 Lyngby
Denmark
e-mail: branner@mat.dtu.dk

HANNE KOCK
Engineering College of Aarhus
Dalgas Avenue 2
DK-8000 Aarhus
Denmark
e-mail: hk@et.aarhus.ih.dk

LISBETH FAJSTRUP
Department of Mathematics and Computer Science
University of Aalborg
Fredrik Bajers Vej 7E
DK-9220 Aalborg Ø
Denmark
e-mail: fajstrup@iesd.auc.dk

LENA M. FINNE
Department of Teacher Education
Åbo Akademi
Postbox 311
FIN-65101 Vasa
Finland

CORNELIA NIEDERDRENK-FELGNER
German Institute for Distance Education Research (DIFF)
Tübingen
Germany

JOSETTE ADDA
Université Lumière (Lyon 2)
86, rue Pasteur
69365 Lyon cedex 07
France

ROSA MARIA SPITALERI
Istituto per le Applicazioni del Calcolo
V.le del Policlinico, 137
00161 Roma
Italy

TERESA SMART
School of Teaching Studies
University of North London
166-220 Holloway Road
London N7 8DB
UK

JANICE GAFFNEY
Department of Applied Mathematics
The University of Adelaide
Adelaide, South Australia 5005
Australia

JUDITH GILL
School of Education
University of South Australia
Holbrooks Road
Underdale, South Australia 5032
Australia

MEGAN CLARK
Centre for Mathematics and Science Education
Victoria University
Wellington
New Zealand

RUI FEN TANG AND QI MING ZHENG
Department of Mathematics
East China Normal University
3663 Zhongshan Road (Northern)
Shanghai, 2000 62
P. R. of China

SHEN QUAN WU
Shanghai Wei Yu Secondary School
1261 Fuxing Road (Middle)
Shanghai
P. R. of China

CARLOS BOSCH AND MARIA TRIGUEROS
ITAM
Depto de Matemaricas
Rio Hondo #1
Col. Tizapan-San Angel
01000 Mexico DF
Mexico

SUSAN F. CHIPMAN
U.S. Office of Naval Research
ONR 342, 800 N. Quincy Street,
Arlington, VA, 22217-5660
USA

New ICMI Studies Series

1. M. Niss (ed.), *Cases of Assessment in Mathematics Education.* 1992
   ISBN 0-7923-2089-1
2. M. Niss (ed.), *Investigations into Assessment in Mathematics Education.* 1992
   ISBN 0-7923-2095-6
3. G. Hanna (ed.), *Towards Gender Equity in Mathematics Education.* 1996
   ISBN 0-7923-3921-5; Pb 0-7923-3922-3

KLUWER ACADEMIC PUBLISHERS – DORDRECHT / BOSTON / LONDON